OTHER TITLES OF INTEREST FROM ST. LUCIE PRESS

The Handbook of Trace Elements

Plant Nutrition Manual

Handbook of Reference Methods for Plant Analysis

An Introduction to Ecological Economics

Seed Ecophysiology of Temperate and Boreal Zone Forest Trees

Privatization of Technology and Information Transfer in Agriculture

Interrelationship Between Insects and Plants

Earthworm Ecology

For more information about these titles call, fax or write:

St. Lucie Press
2000 Corporate Blvd., N.W.
Boca Raton, FL 33431-9868

TEL (561) 994-0555 • (800) 272-7737
FAX (800) 374-3401
E-MAIL information@slpress.com
WEB SITE http://www.slpress.com

S_L^t

HYDROPONICS

A Practical Guide
for the Soilless Grower

HYDROPONICS

A Practical Guide
for the Soilless Grower

J. Benton Jones, Jr.

S^t_L

St. Lucie Press
Boca Raton, Florida

Phone: (561) 994-0555
E-mail: information@slpress.com
Web site: http://www.slpress.com

S_L^t

Published by
St. Lucie Press
2000 Corporate Blvd., N.W.
Boca Raton, FL 33431-9868

Table of Contents

Preface

This book is a revision of an earlier guidebook by the author published in 1983. Significant advances have been made in the past decade in the application of hydroponic/soilless culture methods of growing. Today, over 60,000 acres worldwide are devoted to the greenhouse production of vegetables hydroponically, and it is anticipated that there will be continuing increases in the acreage of greenhouse vegetable production utilizing various hydroponic/soilless methods. Significant advances have been made and will continue to be made in controlling the environment in the greenhouse as well as the breeding of plant cultivars fully adapted to greenhouse conditions. Therefore, growers will need to use and control more precisely the nutrient element supply to plants in order to take full advantage of the environmental controls and plant cultivars being introduced. Many of the systems initially devised for growing plants hydroponically are no longer suitable for this developing technology. Devising hydroponic growing systems for space application, in confined inhospitable environments, and outdoor growing are new challenges that are changing our concepts of how to best utilize limited water resources, fully utilize nutrient elements, and provide for an ideal rooting environment. In many of these new applications, hydroponic/soilless systems must function efficiently without the possibility of failure—a challenge that sometimes reaches beyond our current concepts of how plants grow.

This book is devoted to describing various techniques for growing plants without soil. It begins with the concepts of how plants grow and then describes the requirements necessary to be successful when using various hydroponic and soilless growing methods.

A major focus of the book is the nutritional requirements of plants and how best to prepare and use nutrient solutions to satisfy the nutrient element requirement of plants. Various hydroponic/soilless systems of growing are described in detail, and the advantages and disadvantages of each system are discussed. Numerous nutrient solution formulas are given, and many tables and illustrations are included. A glossary of key terms is also provided.

This book provides much valuable information for the commercial grower, the researcher, the hobbyist, and the student. Students interested in producing nutrient element deficiencies hydroponically in plants are given the necessary instructions to generate such visual symptoms.

The use of trade names and mention of particular products in this book does not imply endorsement of the products named or criticism of similar ones not named, but rather such products are used as examples for illustration purposes.

J. Benton Jones, Jr.

About the Author

J. Benton Jones, Jr. is vice president of Micro-Macro International, an analytical laboratory specializing in the assay of soil, plant tissue, water, food, animal feed, and fertilizer. He is also president of his own consulting firm, Benton Laboratories; vice president of a video production company engaged in producing educational videos; and president of a new company, Hydro-Systems, Inc., which manufactures hydroponic growing systems.

Dr. Jones is Professor Emeritus at the University of Georgia. He retired from the university in 1989 after having completed 21 years of service plus 10 years as Professor of Agronomy at the Ohio Agricultural Research and Development Center, Wooster.

He received his B.S. degree from the University of Illinois in 1952 in agricultural science and a M.S. degree in 1956 and a Ph.D. degree in 1959 in agronomy from the Pennsylvania State University.

Dr. Jones is the author of over 200 scientific articles and 15 book chapters, and he has written four books. He was editor of two international journals, *Communications in Soil Science and Plant Analysis* for 24 years and the *Journal of Plant Nutrition* for 19 years. Dr. Jones is secretary-treasurer of the Soil and Plant Analysis Council, a scientific society which was founded in 1969, and has been active in the Hydroponic Society of America from its inception, serving on its board of directors for five years.

He has traveled extensively with consultancies in the Soviet Union, China, Taiwan, South Korea, Saudi Arabia, Egypt, Costa Rica, Cape Verde, India, Hungary, Kuwait, and Indonesia.

Dr. Jones has received many awards and recognition for his service to the science of soil testing and plant analysis. He is a certified soil and plant scientist under the ARPACS program of the American Society of Agronomy, Fellow of the American Association for the Advancement of Science, Fellow of the American Society of Agronomy, and Fellow of the Soil Science Society of America. An award in his honor, The J. Benton Jones, Jr. Award, established in 1989 by the Soil and Plant Analysis Council, has been given to three international soil scientists, one in each of the years 1991, 1993, and 1995. Dr. Jones received an Honorary Doctor's Degree from the University of Horticulture, Budapest, Hungary, and is a member of three honorary societies, Sigma Xi, Gamma Sigma Delta, and Phi Kappa Phi. He is listed in *Who's Who in America* as well as a number of other similar biographical listings.

Introduction 1

The growing of plants in water or nutrient solution, referred to as *hydroponics* (*hydro* = water, *ponos* = labor, i.e., *working water*), has been practiced for centuries. For example, the ancient Hanging Gardens of Babylon and the floating gardens of the Aztecs in Mexico were hydroponic in nature. In the 1800s, the basic concepts for the hydroponic growing of plants were established by those investigating how plants grew (Steiner, 1985). The soilless culture of plants was then popularized in the 1930s by a California scientist (Gericke, 1929, 1937, 1940).

During the Second World War, the U.S. Army established large hydroponic gardens on several islands in the western Pacific to supply fresh vegetables to troops operating in that area (Eastwood, 1947). Since the 1980s, the hydroponic technique has become of considerable commercial value for vegetable (Elliott, 1989) and flower production (Fynn and Endres, 1994), as today there are over 60,000 acres of greenhouse vegetables being grown hydroponically throughout the world, an acreage that is expected to continue to increase (Jensen, 1995).

Hydroponics for space applications—providing a means of purifying water, maintaining a balance between oxygen and carbon dioxide in space compartments, and supplying food for astronauts—is being intensively researched (Knight, 1989; Schwartzkopf, 1990; Tibbitts, 1991; Brooks, 1992). Hydroponic growing in desert areas of the world (Jensen and Tern, 1971) and in areas such as the polar regions (Tapia, 1985; Rogan and Finnemore, 1992; Sadler, 1995; Budenheim et al., 1995) or other inhospitable regions will become important for providing food and/or a mechanism for waste recycling (Budenheim, 1991, 1993).

Actually, hydroponics is only one form of soilless culture. It refers to a technique in which plant roots are suspended in either a static, continuously aerated nutrient solution or a continuous flow or mist of nutrient solution. Plants growing in an inorganic substance (such as sand, gravel, perlite, or rockwool) or in an organic material (such as sphagnum peat moss or pine bark) and periodically watered with a nutrient solution should be referred to as soilless culture but not necessarily hydroponic. Some may argue with these definitions, as the common conception of hydroponics is that plants are grown without soil, and the required 16 essential elements are provided by means of a nutrient solution that periodically bathes the roots.

Most of the books on hydroponic/soilless culture (see Appendix C) focus on the general culture of plants and the design of the growing system, giving only sketchy details on the rooting bed design and the composition and management of the nutrient solution. Although the methods of solution delivery and plant support media may vary considerably among hydroponic/soilless systems, most have proven to be workable, resulting in reasonably good plant growth. However, there is a difference between a working system and one that is commercially successful. Unfortunately, many workable soilless culture systems are not commercially sound. Most books on hydroponics would lead one to believe that hydroponic/soilless culture methods for plant growing are relatively free of problems since the rooting media and supply of nutrient elements can be controlled. Experience has shown that hydroponic/soilless growing requires careful attention to details and good growing skills. Most hydroponic/soilless growing systems are not easy to manage by the inexperienced and unskilled. Soil growing is more forgiving of errors made by the grower than most hydroponic/soilless growing systems, particularly those that are purely hydroponic.

In 1981, Jensen (1981) listed the advantages and disadvantages of the hydroponic technique for crop production, many of which are still applicable today.

Advantages

1. Crops can be grown where no suitable soil exists or where the soil is contaminated with disease.
2. Labor for tilling, cultivating, fumigating, watering, and other traditional practices is largely eliminated.
3. Maximum yields are possible, making the system economically feasible in high-density and expensive land areas.

4. Conservation of water and nutrients is a feature of all systems. This can lead to a reduction in pollution of land and streams because valuable chemicals need not be lost.
5. Soil-borne plant diseases are more readily eradicated in closed systems which can be totally flooded with an eradicant.
6. More complete control of the environment is generally a feature of the system (i.e., root environment, timely nutrient feeding or irrigation), and in greenhouse-type operations, the light, temperature, humidity, and composition of the air can be manipulated.
7. Water carrying high-soluble salts may be used if done with extreme care. If the soluble salts in the water supply are over 500 ppm, an open system of hydroponics may be used if care is given to frequent leaching of the growing medium to reduce the salt accumulations.
8. The amateur horticulturist can adapt a hydroponic system to home and patio-type gardens, even in high-rise buildings. A hydroponic system can be clean, lightweight, and mechanized.

Disadvantages

1. The original construction cost per acre is great.
2. Trained personnel must direct the growing operation. Knowledge of how plants grow and the principles of nutrition are important.
3. Introduced soil-borne diseases and nematodes may be spread quickly to all beds on the same nutrient tank of a closed system.
4. Most available plant varieties adapted to controlled growing conditions will require research and development.
5. The reaction of the plant to good or poor nutrition is unbelievably fast. The grower must observe the plants every day.

In 1983, Collins and Jensen (1983) prepared another overview of the hydroponic technique of plant production, and more recently, Jensen (1995) discussed probable future hydroponic developments, stating that "the future growth of controlled environment agriculture will depend on the development of production systems that are competitive in terms of costs and returns with open field agriculture" and that "the future of hydroponics appears more positive today than any time over the last 30 years." In a brief review of hydroponic growing activities in Australia, Canada, England, France, and Holland, Brooke (1995b) stated that "today's hydroponic farmer can grow crops safely and in places that were formerly considered too barren to cultivate, such as deserts, the Arctic, and even in space." He concluded, "hydroponic technology spans the globe." Those looking for a brief overview of the common systems of hydroponic growing in use today will find the article by Rorabaugh (1995) helpful.

Proper instruction in the design and workings of a hydroponic/soilless culture system is absolutely essential. Those not familiar with the potential hazards associated with these systems or who fail to understand the chemistry of the nutrient solution required for their proper management and plant nutrition will normally fail to achieve commercial success with most hydroponic/soilless culture systems.

This book describes various systems of hydroponic/soilless growing and those characteristics essential for success. The common procedures for both inorganic and organic media as well as purely hydroponic culture are described, with emphasis on the essentials of the technique. Although the importance of these factors is mentioned in this text, the reader is advised to seek other sources for general information on plant production, such as greenhouse design and construction, environmental control other than the nutrient solution, cultivar selection and cultural plant practices, and pest management.

The technology associated with plant production, hydroponic or otherwise, is rapidly changing, as can be evaluated by reviewing the various bibliographies on hydroponics (Anon., 1984; Gilbert, 1979, 1983, 1984, 1985, 1987, 1992). Those interested in hydroponics must keep abreast of the rapid developments that are occurring by subscribing to and reading periodicals, such as *The Growing Edge*;* by membership and participation in societies, such as the Hydroponic Society of America** and the International Society of Soilless Culture,*** and organizations devoted to the hydroponic/soilless growing of plants; and by becoming acquainted with the books, bulletins, and developing computer, video, and Internet (i.e., e-mail: hydrosoccam@aol.com) sources of hydroponic information.

---

* *The Growing Edge*, P.O. Box 1027, Portland, OR 97339; phone: (503) 757-0027.
** Hydroponic Society of America, 2819 Crow Canyon Road, Suite 218, San Ramon, CA 94583; phone: (510) 743-9605; fax: (510) 743-9302.
*** Secretariat of the International Society of Soilless Culture, P.O. Box 52, Wageningen, The Netherlands.

How Plants Grow

2

The ancient men of knowledge wondered about how plants grow. They concluded that plants obtained nourishment from the soil, calling it a "particular juyce" existent in the soil for use by plants. In the 16th century, van Helmont regarded water as the sole nutrient for plants. He came to this conclusion after conducting the following experiment:

> Growing a willow in a large carefully weighed tub of soil, van Helmont observed at the end of the experiment that only 2 ounces of soil was lost during the period of the experiment, while the willow increased in weight from 5 to 169 pounds. Since only water was added to the soil, he concluded that plant growth was produced solely by water.

Later in the 16th century, John Woodward grew spearmint in various kinds of water and observed that growth increased with the increasing impurity of the water. He concluded that plant growth increased in water that contained increasing amounts of terrestrial matter, because this matter is left behind in the plant as water passes through the plant.

The idea that soil water carried "food" for plants and that plants "live off the soil" dominated the thinking of the times. It was not until the mid to late 18th century that experimenters began to clearly understand how, indeed, plants grow.

A book entitled *The Principle of Agriculture and Vegetation,* published in 1757 by the Edinburgh Society and written by Francis Home, introduced a number of factors believed to be related to plant growth. Further, Home recognized the value of pot experiments and plant analysis as means of determining those factors affecting plant growth. His book attracted considerable attention and led experimenters to explore both the soil and the plant more intensively.

Joseph Priestley's famous experiment in 1775 with an animal and a mint plant enclosed in the same vessel established the fact that plants will "purify" rather than deplete the air, as do animals. His results opened a whole new area of investigation. Twenty-five years later, DeSaussure determined that plants consume carbon dioxide from the air and release oxygen when in the light. Thus, the process that we today call photosynthesis was discovered, although it was not well understood by DeSaussure or others at that time.

At about the same time, and as an extension of earlier observations, the "humus" theory of plant growth was proposed and widely accepted. The concept postulated that plants obtain carbon and essential nutrients (elements) from soil humus. This was probably the first suggestion of what we would today call the "organic gardening" concept of plant growth and well-being. Experiments and observations made by many since then have discounted the basic premise of the "humus theory" that plant health comes only from soil humus sources.

In the middle of the 19th century, an experimenter named Boussingault began to carefully observe plants, measuring their growth and determining their composition as they grew in different types of treated soil. This was the beginning of many experiments demonstrating that the soil could be manipulated through the addition of manures and other chemicals to affect plant growth and yield. However, these observations did not explain why plants responded to changing soil conditions. Then came a famous report in 1840 by Liebig, who stated that plants obtain all their carbon from carbon dioxide in the air. A new era of understanding plants and how they grow emerged. For the first time, it was understood that plants utilize substances in both the soil and the air. Subsequent efforts turned to identifying those substances in soil, or added to soil, that would optimize plant growth in desired directions.

The value and effect of certain chemicals and manures on plant growth took on new meaning. The field experiments conducted by Lawes and Gilbert at Rothamsted (England) led to the concept that substances other than the soil itself can influence plant growth. About this time, the water experiments by Knop and other plant physiologists (a history of how the hydroponic concept was conceived is given by Steiner [1985]) showed conclusively that potassium, magnesium, calcium, iron, and phosphorus, along with sulfur, carbon, nitrogen, hydrogen, and oxygen, are all necessary for plant life. It is interesting to observe that the formula devised by Knop for growing plants in a nutrient solution can still be used successfully today in most hydroponic systems (Table 1).

Table 1   Knop's nutrient solution

Reagent	g/L
Potassium nitrate (KNO$_3$)	0.2
Calcium nitrate [Ca(NO$_3$)$_2 \cdot$ 4H$_2$O]	0.8
Monopotassium phosphate (KH$_2$PO$_4$)	0.2
Magnesium nitrate (MgSO$_4 \cdot$ 7H$_2$O)	0.2
Ferric phosphate (FePO$_4$)	0.1

Keep in mind that the mid-19th century was a time of intense scientific discovery. The investigators named above are but a few of those who made significant discoveries that influenced the thinking and course of scientific biological investigation. Many of the major discoveries of their day centered on biological systems, both plant and animal. Before the turn of the 19th century, the scientific basis of plant growth had been well established, as has been reviewed by Russell (1950). Research had proven conclusively that plants obtained the carbon, hydrogen, and oxygen required for carbohydrate synthesis from carbon dioxide and water by the process later called photosynthesis, that nitrogen was obtained by root absorption of either ammonium and/or nitrate ions (although leguminous plants can supplement with symbiotically fixed nitrogen from the air), and that all the other elements are taken up by plant roots from the soil as ions. This general outline remains the basis for our present understanding of plant growth, although we now know that there are 16 essential elements, and we have extended our knowledge about how these elements function in plants and at what levels they are required to maintain healthy, vigorous growth, as well as how they are absorbed and translocated.

Although there is much that we do know about plants and how they grow, there is still much that we do not understand, particularly about the role of some of the essential elements. Balance, the relationship of one element to another, may be as important as the concentration of any one of the elements in optimizing plant nutrition. There is still considerable uncertainty as to how elements are absorbed by plant roots and how they then move within the plant. Elemental form, whether individual ions or complexes, may be as important for movement and utilization as concentration. For example, chelated iron forms are effective for control of iron deficiency, although ionic iron, either as the ferric (Fe^{3+}) or ferrous (Fe^{2+}) ions, is equally effective but at higher concentrations.

The biologically active portion of an element in the plant, frequently referred to as the *labile* form, may be that portion of the concentration that determines the character of plant growth. The use of tissue tests is partly based on this concept, measuring that portion of the element that is found in the plant sap and then relating this concentration to plant growth.

The science of plant nutrition is attracting considerable attention today as plant physiologists determine how plants utilize the essential elements. In addition, the characteristics of plants can now be genetically manipulated by adding and/or removing traits that alter the ability of the plant to withstand environmental and biological stress and improve product quality (Mohyuddin, 1985; Waterman, 1993–94; Baisden, 1994). With these many advances, all forms of growing, whether hydroponic or otherwise, are now becoming more productive.

Soil and Hydroponics 3

Scientifically speaking, plant growth in all rooting media, including soil, is *hydroponic,* since the elements absorbed by plant roots *must be* in a water-based solution. The concentration and movement of the elements within this solution depend on the nature of the surrounding medium. For example, in soil, the soil solution and its elemental composition are the result of many interacting factors, an ever-changing, dynamic system of complex equilibrium chemistry (Lindsay, 1979) in which the soil, soil microorganisms, and the plant root (Carson, 1974) each play unique and specific roles which alter the *availability* and eventual absorption by the plant root of the elements required for growth (Barber and Bouldin, 1984; Barber, 1995). The complexity of the chemistry of the soil (nutrient) solution is significantly simplified when the support medium is inert, such as in sand, gravel, perlite, and rockwool culture, and becomes even simpler when the plant roots are suspended in a nutrient solution, as is the case in the standing aerated nutrient solution, nutrient film technique, and aeroponic methods of hydroponic growing. Asher and Edwards (1978b) duplicated the soil solution hydroponically in their study of plant nutrition on low-fertility soils.

There are those who consider soil growing as a system that is "out of control," while hydroponics is classed as a system "for control." This would seem at first glance to be a reasonable assessment, although not entirely true in practice. A soil system is indeed difficult to keep in control due to the complex inorganic–organic and biological nature of soil, as well as the interaction of plant roots on soil processes. Plants growing in soil are frequently competitors for the essential elements in the soil solution with other organisms (i.e., bacteria, fungi, etc.) present in the soil. These interactive processes and competition can be minimized in a hydroponic system. Therefore, the grower has the ability to *regulate* the composition of the nutrient solution and, in turn, to control plant growth to a considerable degree. The challenge for the hydroponic grower is the control of the nutrient solution composition, a topic dealt with in some detail in this book.

Those holding the *organic* view of plant growth and development have considerable difficulty in accepting hydroponics as a *natural* system of plant production. Their contention is that unless the elements essential for plants are derived from an *organic* or natural source, plant growth and development are deficient and, therefore, *unnatural*. Scientific proof that such is the case is lacking, although many argue the *natural* point of view with considerable elegance, despite the lack of factual substantiation (Bezdicek, 1984).

A case may be made about purity, which has some degree of scientific validity. For example, it has been demonstrated that in carefully controlled environments, laboratory mice frequently do not do as well when placed in a pure environment (that is, free from substances thought to be harmful and/ or not needed) as compared to those exposed to typically uncontrolled (natural) environments. These experiments suggest that man does not know all there is to know about the growth and well-being of laboratory mice and probably knows less about plants. This observation may have some degree of significance in dealing with a hydroponic system of plant culture when choosing the source of the chemicals, support media, and water used. Within certain limits, a less pure system may be more desirable than one which uses purified ingredients.

There is increasing scientific data which suggest that a number of elements at trace concentrations (at the parts per billion level) can have significant favorable effects on plant growth and development. A new book on the trace elements by Pais and Jones (1997) presents some of the data which justify the suggestion that more than the currently identified 16 essential elements should be included in the nutrient solution for highest plant growth performance. This poses an interesting problem for the soilless culture grower when selecting the method and composition of the nutrient solution to be used. Pure chemicals (analytical or technical grades) used for making the nutrient solution, purified water, and either no support medium or an inert one may not be the best of alternatives. Plant growth, and probably yield, may be considerably better in a culture system that exposes the plant to a more natural environment than one that is "laboratory" sterile. This may also explain the observation that in some instances the *organic system* benefits from the complexities of the natural environment, as compared to a system constituted and controlled by man on the basis of current scientific knowledge.

Even the early researchers, when utilizing the solution culture technique in their studies, recognized that more than the known essential elements must be present in the nutrient solution in order to achieve maximum plant growth. To ensure that all of the elements that might favorably affect plant growth were present in their constituted nutrient solutions, the A–Z Micronutrient Solutions (given in Table 12) were devised. Included in the two-part solutions are 20 elements:

aluminum (Al)	cadmium (Cd)	lead (Pb)	selenium (Se)
arsenic (As)	chromium (Cr)	lithium (Li)	strontium (Sr)
barium (Ba)	cobalt (Co)	mercury (Hg)	tin (Sn)
bismuth (Bi)	fluorine (F)	nickel (Ni)	titanium (Ti)
bromine (Br)	iodine (I)	rubidium (Rb)	vanadium (V)

None of these elements are currently considered essential, except nickel, which has been suggested as essential by several researchers (Brown et al., 1987; Eskew et al., 1984). However, eight of these elements (arsenic, chromium, cobalt, fluorine, iodine, nickel, selenium, and vanadium) are recognized as essential for animals. The A–Z Micronutrient Solutions are not used today, although they do point to the possibility that many of the known elements can be beneficial to plant growth and therefore deserve some attention. The so-called *beneficial elements* are discussed in greater detail in Chapter 6.

Therefore, the hydroponic/soilless media grower should attempt to duplicate in part the role that soil plays in supplying and controlling essential element availability. In some ways, this task is easier and in others more difficult when it must be done almost entirely by means of a chemically made nutrient solution.

In soil, elemental uptake is affected by the movement of the elements in the soil solution and by the growth of plant roots; the various processes involved are discussed by Barber (1995). The movement of elements along with the soil water is called mass flow; it can carry elements to or away from plant roots with water movement. Within the soil solution itself, elements move from regions of high to low concentration by diffusion. Thus, as the ions of elements are absorbed by plant roots from the solution in immediate contact with the root surface, a concentration gradient is formed (a lower ion concentration exists in the soil solution next to the root, as compared to the higher ion concentration away from the root), which provides a mechanism for resupply—ions flow from high to low areas of concentration. The plant also plays a role by root extension (growth) into the soil mass, bringing greater contact between the root surfaces and the soil mass.

Much of the complexity of the root–soil phenomenon is reduced in hydroponic systems, where the plant roots are periodically bathed with a moving nutrient solution which contains most of the essential elements required by the plant. The flow (application) of the nutrient solution acts much like the mass flow behavior in soil systems. Therefore, the impact of diffusion and root extension on elemental availability and root uptake is reduced. It should be noted that in a soil–plant system, only a very small portion of the soil makes contact with plant roots, whereas in most hydroponic systems, plant roots are exposed to almost the full volume of nutrient solution. Such an extensive exposure of rooting surface to the nutrient solution has advantages, but it also poses problems which will be discussed in more detail later.

The Plant Root—Its Roles and Functions

4

Plant roots have two major functions:

- They physically anchor the plant to the growing medium.
- They are the avenue through which water and ions enter into the plant for redistribution to all parts of the plant.

Although the first role given above is important, it is the second role that deserves our attention in this discussion. The book edited by Carson (1974) provides detailed information on plant roots and their many important functions.

Water Uptake

Water is literally *sucked* into the roots due to the loss of water from the plant by a process called *transpiration*, which takes place mainly from leaf surfaces. To understand this process, visualize continuous columns of water from the root surface up to all the atmospherically exposed plant (leaf) surfaces: water enters through the roots, is pulled up through the plant (mainly in the xylem), and is evaporated from the exposed leaf surfaces.

The shape of the plant is determined by its water content, for when water content declines, wilting occurs and the plant begins to lose its shape and begins to droop. There may be conditions where water uptake and movement within the plant are insufficient to keep the plant fully turgid, particularly when the atmospheric demand is high. In order for water to enter the roots, the roots must be fully functional. Temperature, ion content of the water, aeration, pH, etc. are factors that can affect water absorption by roots. The flow of water within the plant itself is fairly complex and is beyond the scope of our discussion here.

As water is pulled into the plant roots, those substances dissolved in the water will also be brought into the plant, although there is a highly selective system that regulates which ions are carried in and which ions are kept out. Therefore, as the amount of water absorbed through plant roots increases, the amount of ions will also increase, even though a regulation system exists. This partially explains why the elemental content of the plant can vary depending on the rate of water uptake. Therefore, atmospheric demand can affect the elemental content of the plant, which can be either beneficial or detrimental.

Ion Uptake

Presently, we do not have a complete, clear understanding of how ions move from the solution surrounding the root into the root and then how these ions are transported to the upper portions of the plant. However, we do know that the absorption of ions by the root is by both a *passive* and an *active* process.

Passive root absorption means that an ion is carried into the root by the passage of water; that is, it is sort of "carried" along in the water taken to the plant. It is believed that the passive mode of transport explains the high concentrations of some ions, such as potassium (K^+), nitrate (NO_3^-), and chloride (Cl^-), found in the leaves and stems of some plants. The controlling factors in passive absorption are the amount of water moving into the plant (which varies with atmospheric demand), the concentration of these ions in the water, and the size of the root system. Passive absorption is not the whole story however, as a process involving chemical selectivity occurs when an ion-bearing solution reaches the outer cells of the root surface.

The outer cell walls of the root surface form an effective barrier to the passage of most ions into the root. Water may move into these cells, but the ions contained in the water will be left behind in the solution surrounding the root. Also, another phenomenon is at work: ions will only move physically from an area of high concentration to one of lower concentration. However, in the case of root cells, the concentration of most ions in the root is higher than that in the water surrounding the root. Therefore, ions should move from the root into the surrounding water; indeed, this can and does happen. The question is how ions move against this concentration gradient and enter the root.

The answer is *active* absorption. It is not entirely clear how this works, but several theories have been proposed to explain active absorption. These theories are based on the nature of cell membranes. Cell membranes function in several ways to control the flow of ions from outside to inside the cell. It is common to talk about "transporting" an ion across the cell membrane and, indeed, this may be what happens. An ion may be complexed with some substance and then "carried" across (or through) the membrane into the cell against the concentration gradient. For the system to work, a carrier must be present and energy expended. As yet, no one has been able to determine what the carrier or carriers are or if indeed they exist. However, the carrier concept helps to explain what is observed in the movement of ions into root cells. The other theory relates to the existence and function of ion pumps rather than specific carriers. For both of these systems to work, energy, which comes from cellular metabolism, is required—a process called *respiration*.

Although we do not know the entire story to exactly explain active absorption, there is general agreement that some type of active system does in fact exist which regulates the movement of ions into the plant root.

There are three things we do know about ion absorption by roots:

1. The plant is able to take up ions selectively even though the outside concentration and ratio of elements may be quite different than those in the plant.
2. Accumulation of ions by the root does occur across a considerable concentration gradient.
3. The absorption of ions by the root requires energy that is generated by cell metabolism.

A unique feature of the active system of ion absorption by plant roots is that it exhibits ion competition, antagonism, and synergism. The competitive effects restrict the absorption of some ions in favor of others. Examples of enhanced uptake relationships include:

- Potassium (K^+) uptake is favored over calcium (Ca^{2+}) and magnesium (Mg^{2+}) uptake.
- Chloride (Cl^-), sulfate (SO_4^{2-}), and phosphate ($H_2PO_4^-$) uptake is stimulated when nitrate (NO_3^-) uptake is strongly depressed.

The rate of absorption is also different for various ions. The monovalent ions (i.e., K^+, Cl^-, NO_3^-) are more rapidly absorbed by roots than the divalent (Ca^{2+}, Mg^{2+}, SO_4^{2-}) ions.

The uptake of certain ions is also enhanced in active uptake. If the nitrate (NO_3^-) ion is the major nitrogen source in the surrounding rooting environment, then there tends to be a balancing effect marked by greater intake of the cations K^+, Ca^{2+}, and Mg^{2+}. If the ammonium (NH_4^+) ion is the major source of nitrogen, then uptake of the cations K^+, Ca^{2+}, and Mg^{2+} is reduced. The presence of NH_4^+ enhances NO_3^- uptake. If Cl^- ions are present in sizable concentrations, NO_3^- uptake is reduced.

These effects of ion competition, antagonism, and synergism are of considerable importance to the hydroponic grower in order to avoid the hazard of creating elemental imbalances in the nutrient solution that will affect plant growth and development. Therefore, the nutrient solution must be properly and carefully balanced initially and then kept in balance during its term of use. Imbalances arising from these ion effects will affect plant growth. Steiner (1980) has discussed in considerable detail his concepts of ion balance when constituting a nutrient solution; his concept is presented in Chapter 7.

Unfortunately, many current systems of nutrient solution management do not effectively provide for the problem of imbalance. This is true not only of systems in which the nutrient solution is managed on the basis of weekly dumping and reconstitution, but also of constant-flow systems.

Indeed, the concept of rapid, constant-flow, low-concentration nutrient solution management is made to look deceptively promising in minimizing the interacting effects of ions in the nutrient solution on absorption and plant nutrition (more about these problems in Chapter 7).

Finally, non-ionic substances, mainly molecules dissolved in the soil water, can also be taken into the root by mass flow. Substances such as amino acids, simple proteins, carbohydrates, urea, etc. can easily enter the plant and contribute to its growth and development.

Physical Characteristics

Root architecture is determined by plant species and the physical environment surrounding the roots. Plant roots grow outward and downward, although most rooting containers are not so designed; the root architecture would suggest a pyramid-shaped container, narrow at the top and wide at the bottom. In soil, it has been observed that feeder roots grow up, not down. This is why plants, particularly trees, do poorly when the soil surface is compacted or physically disturbed. In soil, any root restriction can have a significant impact on plant growth and development due to the reduction in soil–root contact. Root pruning, whether done purposely (bonsai plants) or as the result of natural phenomena, will also affect plant growth and development in soil. Therefore, in most hydroponic/soilless growing systems, roots may extend into a much greater volume of growing area or media than would occur in soil.

Root size, measured in terms of length and extent of branching, as well as color are characteristics that are affected by the nature of the rooting environment. Normally, vigorous plant growth is associated with long, white, and highly branched roots. It is uncertain whether vigorous top growth is a result of vigorous root growth or vice versa.

Tops tend to grow at the expense of roots, with root growth slowing during fruit set. Shoot:root ratios are frequently used to describe the relationship that exists between them, with ratios ranging from as low as 0.5 to a high of 15. Root growth is dependent on the supply of carbohydrates from the tops, and, in turn, the top is dependent on the root for water and essential elements. The loss or restriction of roots can significantly affect top growth. Therefore, it is believed that the goal should be to provide and maintain those conditions which promote good, healthy root development, neither excessive nor restrictive.

The physical characteristics of the root itself play a major role in elemental uptake. The rooting medium and the elements in the medium will determine to a considerable degree root appearance. For example, root hairs will be almost absent on roots exposed to a high concentration (100 mg/L, ppm) of nitrate (NO_3^-). High phosphorus will also reduce root hair development, whereas changing concentrations of the major cations (potassium, calcium, and magnesium) will have little effect on root hair development. Root hairs markedly increase the surface available for ion absorption and also increase the surface contact between roots and the water film around particles in a soilless medium; therefore, their presence can have a marked effect on water and ion uptake.

The question that arises is what constitutes healthy functioning roots for the hydroponic/soilless growing system. The size and extent of root development are not as critical as they are in soil. It has been demonstrated that one functioning root is sufficient to provide all the essential elements required by the plant, with size and extensiveness of the roots being more important for water uptake. Therefore, in most hydroponic systems, root growth and extension are probably far greater than needed, which may actually in turn have a detrimental effect on plant growth and performance.

Aeration

Aeration is another important factor that influences root and plant growth. Oxygen (O_2) is essential for cell growth and activity. If not available in the rooting medium, severe plant injury or death will occur. The energy required for root growth and ion absorption is derived by the process called *respiration,* which requires O_2. Without adequate O_2 to support respiration, water and ion absorption cease and roots die.

Oxygen levels and pore space distribution in the rooting medium will also affect the development of root hairs. Aerobic conditions, with equal distributions of water- and air-occupied pore spaces, promote root growth, including root hairs.

If air exchange between the medium and surrounding atmosphere is impaired by overwatering or the pore space is reduced by compaction, O_2 supply is limited and root growth and function will be adversely affected. As a general rule, if the pore space of a solid medium, such as soil, sand, gravel, or an organic mix containing peat moss or pine bark, is equally occupied by water and air, sufficient O_2 will be present for normal root growth and function (Bruce et al., 1980).

In hydroponic systems where plant roots are growing in a standing solution or a flow of nutrient solution, the grower is faced with a "Catch-22" problem in periods of high temperature. The solubility of O_2 in water is quite low (at 75°F, about 0.004%) and decreases significantly with increasing temperature (see Figure 1). However, since plant respiration and, therefore, O_2 demand increase rapidly with increasing temperature, considerable attention to O_2 supply is required. Therefore, the nutrient solution must be

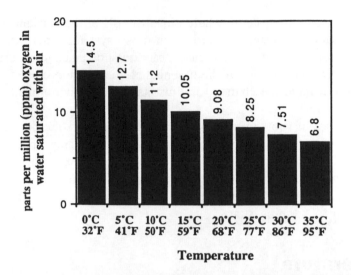

Figure 1   Dissolved oxygen saturation limits for water at sea level pressure and temperature. (Source: Brooke, 1995a.)

kept well aerated by either bubbling air into the solution or by exposing as much of the surface of the solution as possible to air by agitation. One of the significant advantages of the aeroponic system is that plant roots are essentially growing in air and, therefore, are adequately supplied with O_2. Root death, a common problem in most nutrient film technique systems and possibly other growing systems as well, is due in part to lack of adequate aeration within the root mass in the rooting channel.

Root Surface Chemistry

Some plants have the ability to alter the environment immediately around their roots. The most common alteration is a reduction in pH by the emission of hydrogen (H^+) ions. In addition, some plants have the ability to emit substances (such as siderophores) from their roots which enhance ion chelation and uptake. These phenomena have been most commonly observed in species that have the ability to obtain needed iron under adverse conditions and are characteristic of so-called "iron-efficient" plants (Rodriguez de Cianzio, 1991).

This ability of roots to alter their immediate environment may be hampered in hydroponic systems where either the pH of the nutrient solution is being constantly adjusted upward or in those systems where the nutrient solution is not recycled. In such cases, care must be taken to ensure that the proper balance and supply of the essential elements are provided, since the plant may not have the chance to adjust its rooting environment to suit its particular needs.

The impact of roots on a standing aerated nutrient solution system may have an adverse effect on plant growth by either raising or lowering the solution pH, as well as by the introduction of complexing substances into the solution. Therefore, frequent monitoring of the nutrient solution and close observation of plant growth and development can alert the grower to the solution's changing status.

Temperature

Temperature is another important factor which influences root growth, as well as the absorption of water and essential element ions (Nielsen, 1974; Barber and Bouldin, 1984). The optimum root temperature will vary somewhat with plant species, but in general, root temperatures below 68°F (20°C) begin to bring about changes in root growth and behavior. Below optimum temperatures reduce growth and branching and lead to coarser looking root systems. Absorption of both water and ions is also slowed as the permeability of cell membranes and root kinetics are reduced. Translocation in and out of the root is equally slowed with less than optimum root temperatures. When root temperatures are low, plants will wilt on high atmospheric demand days, and elemental deficiencies will appear. Ion absorption of the elements phosphorus, iron, and manganese seems to be more affected by low temperature than most of the other essential elements. It should also be noted that the viscosity of water decreases with decreasing temperature, which in turn affects water movement in and around the plant root.

The maximum root temperature that can be tolerated before significant reduction in root activity occurs is not clearly known. Roots seem to be able to tolerate short periods of high temperature. Roots are fully functional at 86°F (30°C) and probably can withstand temperatures up to 95°F (35°C). However, the current literature is not clear as to the exact limits of the optimum temperature range for best plant growth.

In order to avoid the hazards of either low or high temperatures, the roots and rooting medium should be kept at a temperature between 68 to

86°F (20 to 30°C). Reduced growth and other symptoms of poor nutrition will appear if root temperatures are kept at levels below or above this suggested temperature range.

Root Growth and Plant Performance

A large and extensive root system may not be best for most hydroponic growing systems. Rather than size, active effective roots are needed, since the nutrient solution continuously bathes most of the root system, thereby requiring less surface for absorption to take place. One of the major problems with the nutrient film technique tomato hydroponic system, for example, is the large root mass that develops in the rooting channel, which eventually restricts oxygen (Antkowiak, 1993) and nutrient solution penetration; the end result is a problem called *root death*. Similar extensive root growth occurs with other types of growing systems, particularly with ebb-and-flow systems, where roots frequently grow into the piping that delivers and drains the growing bed of nutrient solution. Similar extensive root growth is obtained with most hydroponic/soilless systems; roots frequently fill bags and blocks of media and sometimes grow beyond the outer walls of bags and containers. The question is whether a large root mass translates into high plant performance. The answer is probably not, if there is more root surface for absorption than needed; in addition, roots require a continuous supply of carbohydrates, which can be used to expand top growth and contribute to fruit yield. Unfortunately, the question as to root size has yet to be adequately addressed.

Hydroponically speaking, a large, ever-expanding root system probably does not necessarily translate into greater top growth and yield and, in fact, may actually have some detrimental effect.

The Essential Elements

5

Through the years, a set of terms has been developed to classify those elements essential for plant growth. This terminology can be confusing and misleading to those unfamiliar with it. Even the experienced can become rattled from time to time.

As with any body of knowledge, an accepted jargon develops which is understood well only by those actively engaged in the field. One of the commonly misused terms when referring to the essential metallic elements, such as copper, iron, zinc, etc., is *mineral*. The strict definition of mineral refers to a compound of elements and not a single element. Yet, *mineral nutrition* is used when referring to plant elemental nutrition. This phrase occasionally appears in conjunction with other words, such as *plant mineral nutrition* or *plant nutrition*—all of which refer to the essential elements and their influence on plants.

Another commonly misused and misunderstood word is *nutrient*, referring again to an essential element. It is becoming increasingly common to combine the words *nutrient* and *element* to mean an essential element. Therefore, elements like nitrogen, phosphorus, potassium, etc. are called *nutrient elements*. Unfortunately, no one has suggested an appropriate terminology when talking about the essential elements; thus, the literature on plant nutrition contains a mixture of these words. In this book, *essential element* and *element* are used in place of *nutrient element* and *nutrient*.

The early plant investigators developed a set of terms to classify the 16 elements identified as essential for plants; these terms have undergone changes in recent times. Initially, the major elements, so named because they are found in sizable quantities in plant tissues, included carbon, hydrogen, nitrogen, oxygen, phosphorus, and potassium, Those elements found in smaller quantities, at first called the *minor elements* or sometimes *trace elements*, are boron, chlorine, copper, iron, manganese, molybdenum, and zinc. More recently, these elements have been renamed *micronutrients*, a term which better fits the comparative ratios between the major elements found in sizable concentrations and the micronutrients found at lower concentrations in plant tissues. The average concentration of the essential elements in plants is given in Table 2 using the data by Epstein (1972).

More recently, Ames and Johnson (1986) listed the major elements by their internal concentrations found in higher plants, as shown in Table 3.

Another term that has been used to designate some of the micronutrients is *heavy metals*, which refers to those elements that have relatively high atomic weights. One definition is "those metals which have a density greater

Table 2   Average concentrations of mineral nutrients in plant dry matter that are sufficient for adequate growth

Element	Symbol	μmol/g Dry Weight	mg/kg (ppm)	%	Relative Number of Atoms
Molybdenum	Mo	0.001	0.1	—	1
Copper	Cu	0.10	6	—	100
Zinc	Zn	0.30	20	—	300
Manganese	Mn	1.0	50	—	1,000
Iron	Fe	2.0	100	—	2,000
Boron	B	2.0	20	—	2,000
Chlorine	Cl	3.0	100	—	3,000
Sulfur	S	30	—	0.1	30,000
Phosphorus	P	60	—	0.2	60,000
Magnesium	Mg	80	—	0.2	80,000
Calcium	Ca	125	—	0.5	125,000
Potassium	K	250	—	1.0	250,000
Nitrogen	N	1,000	—	1.5	1,000,000

Source: Epstein, 1972.

Table 3   Internal concentrations of essential elements in higher plants

Element	Concentration in Dry Tissue	
	ppm	*%*
Major Elements		
Carbon (C)	450,000	45
Oxygen (O)	450,000	45
Hydrogen (H)	60,000	6
Nitrogen (N)	15,000	1.5
Potassium (K)	10,000	1.0
Calcium (Ca)	5,000	0.5
Magnesium (Mg)	2,000	0.2
Phosphorus (P)	2,000	0.2
Sulfur (S)	1,000	0.1
Micronutrients		
Chlorine (Cl)	100	0.01
Iron (Fe)	100	0.01
Manganese (Mn)	50	0.005
Boron (B)	20	0.002
Zinc (Zn)	20	0.002
Copper (Cu)	6	0.0006
Molybdenum (Mo)	0.1	0.00001

Source: Ames and Johnson, 1986.

than 5 mg/cm^3, with elements such as cadmium, cobalt, copper, iron, lead, molybdenum, nickel, and zinc" being considered as heavy metals (Ashworth, 1991).

Unfortunately, three of the essential elements, calcium, magnesium, and sulfur, were initially named *secondary elements*. These so-called secondary elements should be classed as major elements, and they are referred to as such in this text.

Another category that has begun to make its way into the plant nutrition literature is the so-called *beneficial elements*. Discussion of these elements will be deferred for consideration in Chapter 6.

Another category is the *trace elements*, which includes those elements found in plants at very low levels (<1 ppm) but not identified as either essential or beneficial. Some of these trace elements are found in the A–Z Micronutrient Solutions (see Table 12). Another set of elements being investigated is the rare-earth elements, as their presence in plants seems to have stimulatory effects (Pais and Jones, 1997).

The word *available* has developed a specific connotation in plant nutrition parlance. It refers to that form of an element that can be absorbed by the plant. Although its use has been more closely allied with soil growing, it has inappropriately appeared in the hydroponic literature. In order for an element to be taken into the plant, it must be in a soluble form in water solution surrounding the roots. The available form for most elements in solution is as an ion. It should be pointed out, however, that some molecular forms of the elements can also be absorbed. For example, the molecule urea (a soluble form of nitrogen) and some chelated complexes (such as FeEDTA) can be absorbed by the plant root. The elemental form will be discussed in more detail later.

The criteria for essentiality were established by two University of California plant physiologists in a paper published in 1939; Arnon and Stout (1939) described three requirements that an element had to meet in order to be considered essential for plants:

1. Omission of the element in question must result in abnormal growth, failure to complete the life cycle, or premature death of the plant.
2. The element must be specific and not replaceable by another.
3. The element must exert its effect directly on growth or metabolism and not by some indirect effect, such as by antagonizing another element present at a toxic level.

Some plant physiologists feel that the criteria established by Arnon and Stout may have inadvertently fixed the number of essential elements at the current 16 and that for the foreseeable future no additional elements will be found which meet these criteria for essentiality. The 16 essential elements, the discoverer of the element, the discoverer of essentiality, and the date of discovery are given in Table 4; the 16 essential elements, the form utilized by plants, and their function in plants are given in Table 5.

In the case of higher animals, 25 elements have been recognized as essential; of the 16 elements essential for plants, only boron is not required by animals. The nine elements required by animals but not plants are arsenic, chromium, cobalt, fluorine, iodine, nickel, selenium, silicon, and vanadium. A list of the elements essential for both plants and animals is given in Table 6, and their position in the Periodic Table is shown in Figure 2.

Table 4   Discoverer of element and discoverer of essentiality for the essential elements

Element	Discoverer	Year	Discoverer of Essentiality	Year
C	*	*	DeSaussure	1804
H	Cavendish	1766	DeSaussure	1804
O	Priestley	1774	DeSaussure	1804
N	Rutherford	1772	DeSaussure	1804
P	Brand	1772	Ville	1860
S	*	*	von Sachs, Knop	1865
K	Davy	1807	von Sachs, Knop	1860
Ca	Davy	1807	von Sachs, Knop	1860
Mg	Davy	1808	von Sachs, Knop	1860
Fe	*	*	von Sachs, Knop	1860
Mn	Scheele	1774	McHargue	1922
Cu	*	*	Sommer	1931
			Lipman and MacKinnon	1931
Zn	*	*	Sommer and Lipman	1926
Mo	Hzelm	1782	Arnon and Stout	1939
B	Gay Lussac and Thenard	1808	Sommer and Lipman	1926
Cl	Scheele	1774	Stout	1954

* Element known since ancient times.

Source: Glass, 1989.

Some plant physiologists feel that it is only a matter of time before the essentiality of cobalt, nickel, silicon, and vanadium will be added to the current list of 16 essential plant elements and would recommend that these elements be added to the rooting medium to ensure best plant growth. A more detailed discussion of these elements and others classified as beneficial elements can be found in Chapter 6.

The Major Elements

Nine of the 16 essential elements are classified as *major elements*: carbon, hydrogen, oxygen, nitrogen, phosphorus, potassium, calcium, magnesium, and sulfur. Since the first three are obtained from carbon dioxide in the air and water from the rooting media and then combined by photosynthesis to form carbohydrates:

carbon dioxide (CO_2) + water ($6H_2O$)

↓

(in the presence of light and chlorophyll)

↓

carbohydrate ($C_6H_{12}O_6$) + oxygen (O_2)

they are not normally discussed in any detail as unique to hydroponic/soilless growing systems.

Carbon, hydrogen, and oxygen represent about 90 to 95% of the dry weight of plants and are indeed major constituents of plants. The remaining six major elements (nitrogen, phosphorus, potassium, calcium, magnesium, and sulfur) are more important to the hydroponic/soilless culture grower,

Table 5   List of essential elements for plants by form utilized and biochemical function

Essential Elements	Form Utilized	Biochemical Function in Plants
C, H, O, N, S	In the form of CO_2, HCO_3^-, H_2O, O_2, NO_3^-, NH_4^+, N_2, SO_4^{2-}, SO_2. The ions from the soil solution, the gases from the atmosphere.	Major constituents of organic material. Essential elements of atomic groups which are involved in enzymatic processes. Assimilation by oxidation-reduction reactions.
P, B	In the form of phosphates, boric acid, or borate from the soil solution.	Esterification with native alcohol groups in plants. The phosphate esters are involved in energy transfer reactions.
K, Mg, Ca, Mn, Cl	In the form of ions from the soil solution.	Non-specific functions establishing osmotic potentials. More specific reactions by which the conformation of the enzyme protein is brought into optimum status (enzyme activation). Bridging of reaction partners. Balancing indiffusible and diffusible anions.
Fe, Cu, Zn, Mo	In the form of ions or chelates from the soil solution.	Present predominantly in a chelated form incorporated in prosthetic groups. Enable electron transport by valency change.

Source: Mengel and Kirkby, 1987.

Table 6   Essential elements for plants and animals

Life Form	Major Elements	Micronutrients
Plants and animals	Calcium (Ca) Carbon (C) Hydrogen (H) Magnesium (Mg) Nitrogen (N) Oxygen (O) Phosphorus (P) Potassium (K) Sulfur (S)	Chlorine (Cl) Copper (Cu) Iron (Fe) Manganese (Mn) Molybdenum (Mo) Zinc (Zn)
Plants only	Sodium (Na)	Boron (B)
Animals only		Arsenic (As) Chromium (Cr) Cobalt (Co) Fluorine (F) Iodine (I) Nickel (Ni)* Selenium (Se) Vanadium (V)

* Nickel has been suggested as being an essential micronutrient for plants (Eskew et al., 1984; Brown et al., 1987).

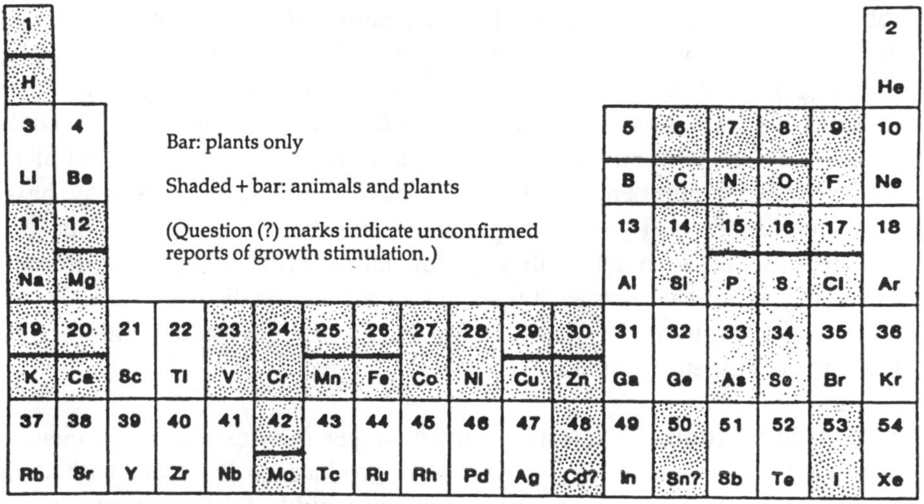

Figure 2   Part of the Periodic Table showing those elements essential for plants and animals.

since these elements must be present in the nutrient solution or added to a soilless medium in sufficient quantity and in the proper balance to meet the crop requirement. Most of the remaining 5 to 10% of the dry weight of plants is made up of these six elements.

Nitrogen

Content in Plants—Nitrogen (N) leaf content of normal plants will vary from a low of 2 to 3% of the dry matter up to 4 to 5%, depending primarily on plant species. The N requirement of plants as a percentage of dry weight is highest during the early stages of growth and then decreases with age. However, the total N requirement will increase substantially up to the reproductive stage of growth and then decline sharply.

Function—Nitrogen is a major constituent of amino acids and proteins, which play essential roles in plant growth and development. Of all the essential elements, N probably has a greater total influence on plant growth than most of the other essential elements, as its deficiency or excess markedly affects plant growth, fruit yield, and quality.

Deficiency Symptoms—Nitrogen deficiency appears as a lightening of the normal green color associated with a healthy appearance. Since N is a mobile element in the plant, the first symptoms of N deficiency appear in the older leaves, which become light green; as the deficiency intensifies, they turn yellow and eventually die. Deficiency symptoms may develop quickly but can just as quickly be corrected by adding some form of available N to the growing medium at a concentration sufficient for normal plant growth and development to resume. Periods of inadequate N may have considerable effect on growth, appearance, and final yield; they are particularly severe on the plant if they occur during critical stages of growth. Deficiency is best confirmed by means of a plant analysis for total N, by tissue tests for nitrate (NO_3), and more recently by the use of a chlorophyll meter measurement (Wood et al., 1993).

Excess Symptoms—There is as much danger in N excess as deficiency, particularly for fruiting crops. Excess N produces lush plants with dark green foliage, but such plants are quite susceptible to disease and insect attack and have greater sensitivity to changing growing conditions. Excess

N in fruiting crops not only impairs blossom set and fruit development but also reduces fruit quality. Excess N frequently does more permanent damage to the plant than does N deficiency.

Forms of Utilization—In most soilless culture systems, proper control of N relates to both concentration and form of the element in the nutrient solution. Most nutrient solution formulas call for a balance between the two common ionic forms of N, nitrate (NO_3^-), and ammonium (NH_4^+), which in turn provides some degree of pH control.

Ammonium versus Nitrate—Experience has shown that the percentage of ammonium ions in the nutrient solution should not exceed 50% of the total N concentration; the better ratio is 75% nitrate to 25% ammonium ions. If ammonium is the major source of N in the nutrient solution, ammonium toxicity can result. However, some ammonium may be desirable, as experiments have shown that the presence of ammonium in the nutrient solution stimulates the uptake of nitrate. It has been shown that as little as 5% of the total N in solution as ammonium in a flowing nutrient solution system is sufficient. A higher percentage will be needed for aerated standing nutrient solution systems; up to 25% of the total N should be ammonium in order to obtain the same stimulating effect on nitrate uptake. Variations of these suggested percentages may be required, depending on plant species, stage of plant growth, nutrient solution flow rate, etc.

Another factor that needs to be considered when selecting the proper ratio of ammonium to nitrate in the nutrient solution is plant species. Fruiting plants, such as tomato and pepper, are particularly sensitive to ammonium nutrition. When ammonium is present in the nutrient solution during flower and fruit initiation, fruit yields are lowered, and a physiological disorder in the fruits, called *blossom-end-rot,* is very likely to occur. Therefore, ammonium may be included in the nutrient solution during the early vegetative growth period but should then be excluded from flower initiation to the end of the growth cycle.

Effect on Roots and Elemental Uptake—It has recently been observed that the N concentration in the nutrient solution can influence the character of root growth. As the nitrate concentration increases, the number and length of root hairs decrease. Concentrations of the other major elements, phosphorus, potassium, calcium, and magnesium, have no similar effect. Even a change in the ammonium content of the nutrient solution has no effect on root hairs. However, roots exposed to high concentrations of ammonium or nutrient solutions, where the major source of N is ammonium, will have coarser looking roots, with little branching or fine root structure. Root growth and its influence on plant growth are discussed in Chapter 4.

Ammonium Toxicity—Ammonium can be toxic to plants when it is the major source of N, resulting in slowed growth and development. Lesions develop on the stem and leaves, and leaves develop a cupping appearance. The vascular tissue then begins to deteriorate (ammonium interferes with calcium function, as calcium is required for maintaining cell wall integrity), causing the plant to wilt on high atmospheric demand days. Ammonium toxicity may eventually result in the death of the plant. If the stem of the affected plant is cut through just above the root line, a darkened ring of decayed vascular tissue is usually clearly visible. Some diseases produce the same symptoms; therefore, careful examination and testing may be needed to determine whether an organism present in the tissue is causing the decay or ammonium toxicity is indeed the cause.

Content in Nutrient Solution—Most formulas call for the total N concentration in the nutrient solution to range from 100 to 200 mg/L (ppm), with the ratio of nitrate to ammonium ions at about 3 or 4 to 1.

Control of Content—Instructions frequently call for the total N concentration in the nutrient solution to start at a lower level (<100 mg/L, ppm), which is then increased as the growing crop matures. This is a common practice in the case of fruiting crops when control of the N supply is directed to minimize excessive vegetative growth and to promote fruit initiation and development. Since N is a key essential element affecting plant growth and quality, careful control of its supply to the plant is extremely important. In soilless growing systems, success or failure hinges to a considerable degree on how well this element is managed.

Reagent Sources—The common sources for nitrate-N are calcium nitrate [$Ca(NO_3)_2 \cdot 4H_2O$], potassium nitrate (KNO_3), and nitric acid (HNO_3); for both ammonium and nitrate, ammonium nitrate (NH_4NO_3); and for ammonium only, ammonium sulfate [$(NH_4)SO_4$] and ammonium mono- or dihydrogen phosphate [$(NH_4)_2HPO_4$ or $NH_4H_2PO_4$, respectively].

Phosphorus

Content in Plants—Phosphorus (P) in plants ranges from 0.2 to 0.5% of the dry matter. The P concentration in young plants is frequently quite high (0.5 to 1%) and slowly declines with plant age; like nitrogen, however, total P uptake increases up to the period of fruit set and then drops off sharply.

Function—Biochemically, P plays a key role in the plant's energy transfer system; thus, P deficiency slows growth considerably.

Deficiency Symptoms—The first symptom of P deficiency is slowed growth. As the deficiency intensifies, the older leaves develop a deep-purple coloring. A similar discoloration can also be brought on by cool temperatures, either in the rooting media or surrounding atmosphere. Since P uptake by plants is somewhat temperature affected, a moderate P deficiency with accompanying symptoms may be induced by cool temperatures; the deficiency symptoms disappear when temperatures return to the normal range. Phosphorus deficiency can be easily detected by means of a plant analysis; deficiency occurs when the leaf concentration is less than 0.2% in most plants. A tissue test for P may also be used to confirm a suspected deficiency.

Excess Symptoms—Until recently, P excess has not been thought of as a common problem. However, recent studies have found that excess P can occur and will significantly affect plant growth. There is accumulating evidence that if the P content of the plant exceeds 1% of its dry weight, P toxicity will result. Phosphorus toxicity is most likely an indirect effect in as much as it affects the normal functions of other elements, such as iron, manganese, and zinc; the interference with zinc is the most likely to occur first. The likelihood of excess seems to be a problem more closely associated with soilless culture than growing in soil, although any form of container growing is subject to the hazard of P overfertilization. In some types of media culture, an initial application of P fertilizer may be sufficient to satisfy the crop requirement without the need for further additions. Phosphorus overfertilization occurs most frequently when the grower uses a general-purpose fertilizer containing P when the only element(s) needed is one (or two) of the other major elements, such as nitrogen and/or potassium.

Content in Nutrient Solution—Most nutrient solution formulas call for a P concentration in solution between 30 to 50 mg/L (ppm). In a continuously flowing nutrient solution, the P level can be significantly less (1 to 2 ppm) and still maintain plant sufficiency. The form of P in solution is either the mono- or dihydrogen phosphate (HPO_4^{2-} or $H_2PO_4^-$, respectively) anions; the particular dominant anion form is a function of the pH of the nutrient solution.

Reagent Sources—Ammonium and potassium, as either the mono- or dihydrogen phosphate [$(NH_4)_2HPO_4$, $NH_4H_2PO_4$; K_2HPO_4, KH_2PO_4, respectively], are the more common reagents used to supply P in nutrient solutions. More recently, phosphoric acid (H_3PO_4) has come into increasing use as a P source when the addition of either ammonium or potassium is not desired.

Potassium

Content in Plants—Potassium (K) content in the plant ranges from 1.25 to 3% of the dry matter. Like nitrogen and phosphorus, the K concentration in the plant is initially high (>5%) and then declines with age. The uptake of K is substantial during vegetative growth and declines rapidly after fruiting. In most fruiting crops, such as tomato, the demand for K by the developing fruit is high. Therefore, plants without adequate K during this critical stage of development will produce fruit of significantly reduced quality. Long-term post-harvest quality of fruits and flowers can be affected by K, requiring higher levels of plant K than that required for plant sufficiency. Since K is mobile in the plant, it can move rapidly from the older tissue to the younger, such as developing fruit. Therefore, a K deficiency can quickly result in visual symptoms in the older plant tissue.

Function—It is uncertain what specific role K plays in the plant, although most plant physiologists believe it is essential for maintaining the proper ion balance in the plant and is also important for carbohydrate synthesis and movement.

Deficiency Symptoms—The initial symptom of K deficiency is slowed growth. As the severity of the deficiency increases, the lower leaves will develop a marginal chlorosis. Potassium deficiency symptoms have been described as a leaf *scorch*, where the leaf has the appearance of having been "burned" along its edges.

Balance Among Cations—There is a critical balance between the K, calcium (Ca), and magnesium (Mg) cations, and when not in balance, plant stress occurs. When K is high in comparison to Ca or Mg, the first likely symptom to occur is Mg deficiency. In some instances, the imbalance can induce a Ca deficiency. An imbalance between these three cations is usually the result of excessive K fertilization, as K is more readily absorbed and transported in the plant than either Ca or Mg. This antagonism is greater between K and Mg than between K and Ca. Despite these differences, care must be taken to ensure that the proper balance between K and both Ca and Mg is maintained so that an induced deficiency of either element does not occur (see Steiner, 1980). For best growth and fruit production for tomato, for example, the content of K and Ca in recently mature leaves should be equal.

Content in Nutrient Solution—Most hydroponic formulas call for the K concentration in the nutrient solution to be around 200 mg/L (ppm), in the form of the potassium (K^+) cation.

Reagent Sources—The common reagents for supplying K are potassium nitrate (KNO_3), potassium sulfate (K_2SO_4), or potassium chloride (KCl).

Calcium

Content in Plants—Calcium (Ca) content in plants ranges considerably, from 0.5 to 2% of the dry weight; the critical value depends on the plant species. In some species, relatively little soluble or what may be referred to as *free* Ca is found in plant tissue, with most of the Ca existing as crystals of calcium oxalate or as precipitates of either calcium carbonate and/or phosphate. It has been suggested that the Ca requirement for plants is very low (about 0.08%), comparable to that of a micronutrient, with higher concentrations required to detoxify the presence of other cations, particularly the heavy metals, such as manganese, copper, and zinc (Wallace, 1971).

Calcium uptake rate is less than that for potassium but remains fairly constant during the life of the plant. The rate of Ca uptake is also dependent on the counter-ions in solution; it is highest when nitrate (NO_3^-) is present in the nutrient solution. It is generally believed that Ca uptake is by passive means and that its movement within the plant is by means of the transpiration stream. However, with maturity, Ca movement becomes restricted, and the influx into leaves and developing fruit slows, resulting in blossom-end-rot, a physiological breakdown of the tissue at the blossom end of the fruit.

Function—Calcium is a major constituent of cell walls; it maintains membrane integrity, which is probably its major, if not its only, significant function in plants.

Deficiency Symptoms—Calcium deficiency or excess occurs in nutrient solutions most commonly as the result of an imbalance with the potassium (K^+) and magnesium (Mg^{2+}) cations. In nutrient solution formulas with the ammonium ion (NH_4^+) as the major source of nitrogen, this ion may act like potassium and become a part of the cation balance and, therefore, affect the uptake of Ca from the nutrient solution.

One of the results of ammonium toxicity is the breakdown of the vascular tissue in the main stem of the plant, which affects cell wall integrity. This condition is thought to be the result of a Ca deficiency induced by a cation imbalance in the nutrient solution.

Calcium deficiency primarily affects leaf appearance, changing the shape of the leaf and turning the tip brown or black. New emerging leaves will have a torn appearance as the margins stick together, tearing the leaf along its margins as it expands. Some leaves may never fully expand to normal size and shape when Ca is deficient. One of the major effects of Ca deficiency is blossom-end-rot of the developing fruit, which is the result of cellular breakdown at this growing point.

Excess Symptoms—Calcium excess is not a common occurrence, although a high Ca concentration may affect the relationship between the major cations potassium and magnesium. Calcium excess may induce either potassium or magnesium deficiency, the latter most likely to be affected first.

Content in Nutrient Solution—The concentration of Ca required in most nutrient solution formulas is around 200 mg/L (ppm). Calcium exists in the nutrient solution as the divalent cation, Ca^{2+}.

Reagent Sources—The major reagent source is calcium nitrate [$Ca(NO_3)_2 \cdot 4H_2O$]. Calcium sulfate ($CaSO_4$) can be used only as a supplementary source of Ca due to its low water solubility. Also, calcium chloride ($CaCl_2$) may be used to a limited degree at rates designated to keep the chloride concentration less than 100 mg/L (ppm). Natural waters may contain a substantial quantity of Ca, as much as 100 mg/L (ppm), sufficient to meet or provide a substantial portion of the nutrient formula requirements. Therefore, when preparing a nutrient solution using such water (frequently referred to as *hard water*), the quantity of Ca contributed by the water should be determined so that the proper Ca concentration in solution is not exceeded.

Magnesium

Content in Plants—Magnesium (Mg) content in the plant will range from 0.2 to 0.5%. The frequency of Mg deficiency may equal that of nitrogen as the result of effects due to improper balance between the other major cations, calcium (Ca^{2+}), potassium (K^+), and ammonium (NH_4^+). In addition, some plant species are more sensitive to Mg than others. Magnesium uptake, like calcium, tends to remain fairly constant with time, but it differs from calcium in that it is more mobile in the plant.

Function—Magnesium is a major constituent of the chlorophyll molecule, the substance in which photosynthesis takes place. Magnesium is also an enzyme activator for a number of important energy transfer processes. Therefore, a deficiency will have serious impact on plant growth and development.

Deficiency Symptoms—Magnesium deficiency symptoms are quite distinct as an interveinal chlorosis, which appears first on the older leaves. Once a Mg deficiency occurs, it is very difficult to correct, particularly if the deficiency occurs during the mid-point in the growing season. In those plant species that have a high Mg requirement, the deficiency may be triggered by various types of environmental and physiological stress. Deficiency can result from an imbalance between potassium and Mg or NH_4 and Mg; calcium may also play a role. An interesting side effect of Mg deficiency is a possible increase in susceptibility to fungus disease infestation as well as the incidence of blossom-end-rot.

Excess Symptoms—Under normal conditions, Mg excess is not likely to occur. However, some investigators suggest that Mg concentrations in the nutrient solution, as well as the plant, should not exceed that of calcium in order to maintain the proper cation balance for best plant growth and development.

Content in Nutrient Solution—Most hydroponic formulas call for Mg at a concentration around 50 mg/L (ppm) in the nutrient solution. Magnesium is present in the nutrient solution as the divalent cation Mg^{2+}.

Reagent Sources—The primary reagent for Mg is magnesium sulfate ($MgSO_4 \cdot 7H_2O$). Natural waters may contain a substantial quantity of Mg, as much as 50 mg/L (ppm). Therefore, when preparing a nutrient solution, the quantity of Mg contributed by the water should be determined so that the proper Mg concentration in the nutrient solution is not exceeded.

Sulfur

Content in Plants—Sulfur (S) content ranges from 0.15 to 0.5% of the plant dry matter. Some authorities feel that the relationship of S to nitrogen (N) is far more important than S concentration alone. Therefore, the N/S ratio might be the better measure of S sufficiency in the plant than total S alone. Equally important may be the amount of sulfate (SO_4)-S present in the plant. Some plant physiologists have suggested the use of the ratio of SO_4-S to total S as the best indicator of sufficiency for this element. Therefore, the literature at the present time is confusing as to the best measure of S sufficiency in plants.

Function—Sulfur is a constituent of two amino acids, cystine and thiamine, which play essential roles in the plant. Plants in the Leguminosae and Cruciferae families have higher requirements for S than most others. They contain a number of S compounds which are easily recognized by their odor and flavor.

Deficiency Symptoms—Sulfur deficiency symptoms are quite similar to those of N deficiency and therefore can confuse even those most expert in plant nutrition evaluation. In general, S deficiency symptoms appear as an overall loss of green color in the plant rather than a loss of primary color in the older leaves, which is the typical N deficiency symptom. It may be necessary, and is probably best, to rely on a plant analysis to confirm a possible S and/or N deficiency problem, rather than relying on visual symptoms alone.

Content in Nutrient Solution—Most hydroponic formulas call for a S concentration of 50 mg/L (ppm) in solution as the sulfate (SO_4^{2-}) anion.

Reagent Sources—The sulfate salts of potassium, magnesium, and ammonium [K_2SO_4, $MgSO_4 \cdot 7H_2O$, $(NH_4)_2SO_4$, respectively] are frequently selected as one of the major sources for nitrogen, potassium, or magnesium, which automatically adds S to the nutrient solution. Little is known about S excess, including whether it can occur and in what form. Evidently plants can tolerate a high concentration of the SO_4^{2-} anion in a nutrient solution without harm to the plant.

The Micronutrients

Plants require considerably smaller concentrations of the micronutrients than the major elements, but the micronutrients are as critically essential as the major elements. The optimum concentrations for the micronutrients are typically in the range of 1/10,000th of the concentration range required for the major elements (see Tables 2 and 3). The micronutrients, as a group, are far more critical in terms of their control and management than some of the major elements, particularly in soilless culture systems. In the case of several of the elements, the required range is quite narrow. Departure from this narrow range results in either deficiency or toxicity when below or above, respectively, the desired concentration range in the rooting media. Deficiency or toxicity symptoms are usually difficult to evaluate visually and therefore require an analysis of the plant for confirmation.

A deficiency of a micronutrient can usually be corrected easily and quickly, but when dealing with excesses or toxicities, correction can be difficult, if not impossible. If toxicity occurs, the grower may well have to start a new crop. Therefore, great care must be taken to ensure that neither insufficient nor excess concentrations of the micronutrients are introduced into the rooting media, either initially or during the growing season.

The availability of some of the micronutrients, particularly iron and zinc, can be significantly changed with a change in pH or with a change in the concentration of one of the major elements, such as phosphorus. Therefore, proper control of the pH and concentration of the major elements in a nutrient solution is critical.

There may be sufficient concentration of some of the micronutrients in the natural environment (i.e., in the water used to make a nutrient solution, the inorganic or organic rooting media, or from contact with piping, storage tanks, etc.) to preclude the requirement to supply a micronutrient by addition. Therefore, it is best to analyze a prepared nutrient solution after constituting it and after contact with its environment to determine its micronutrient content. In addition, careful monitoring of the rooting media and plants will ensure that the micronutrient requirement is being satisfied but not exceeded. Such testing and evaluation procedures are discussed in greater detail in Chapter 12.

Boron

Content in Plants—The sufficiency range for boron (B) is from 10 to 50 mg/kg (ppm) of the dry weight, with the critical values being closer to either the lower or upper concentration of the sufficiency range, depending on the plant species.

Function—The exact function of B in the plant is not clearly known, although there is considerable evidence that it is important in carbohydrate synthesis and transport, pollen growth and development, and cellular activities (division, differentiation, maturation, respiration, growth, etc.).

Deficiency Symptoms—Plants deficient in B exhibit various visual symptoms of deficiency; the first is slowed and stunted new growth, followed by a general stunting of the whole plant. Fruit development will be slow or non-existent, depending on the severity of the deficiency. Fruit quality will be impaired when B is inadequately supplied. When the deficiency is severe, the growing tip of both tops and roots will die.

Excess Symptoms—Because B accumulates in the leaf margins, an early symptom of excess B is discoloration and eventual death of the leaf margins. Normally, discoloration along the whole length of the leaf distinguishes B excess from calcium deficiency, where just the leaf tip and margin at the tip turn brown and die. Boron toxicity can easily result from excess B in the nutrient solution or from B found in natural waters. The B level in the plant should be closely monitored by plant analysis and by care in making the nutrient solution and evaluating the quality of water used.

Content in Nutrient Solution—Hydroponic formulas usually call for a B concentration of about 0.3 mg/L (ppm) in the nutrient solution; the borate (BO_3^{3-}) anion and molecular boric acid (H_3BO_3) are the forms found in solution and utilized by plants.

Reagent Sources—Boric acid (H_3BO_3) is the common reagent source.

Chlorine

Content in Plants—Leaf content of chlorine (Cl) ranges from low parts per million levels (20 mg/kg) in the dry matter to percent (0.15%) concentrations. Levels in excess of 1% would be excessive for most crops.

Function—Relatively little is known about Cl function, but plants tend to wilt easily when a deficiency exists, and some crop species, particularly small grains, become susceptible to various fungus diseases when Cl levels in the plant are low. The critical concentration range is thought to be between 70 to 100 mg/kg (ppm) in dry matter, but higher (1000 mg/kg) for the small grains.

Deficiency/Excess—Since the chloride (Cl^-) anion is ever-present in the environment, deficiencies are not likely to occur, except under special circumstances. There is far greater danger in excesses of Cl resulting from exposure of plants to salt-affected Cl-based environments. Symptoms of Cl toxicity include burning of the leaf tips or margins and premature yellowing and loss of leaves.

Content in Nutrient Solution—Because Cl is a common contaminant in water and reagents used to prepare the nutrient solution, this element does not normally have to be added. Care should be taken to avoid adding sizable quantities of Cl to the nutrient solution by using reagents such as potassium or calcium chloride (KCl or $CaCl_2$, respectively). If present in high concentration in the nutrient solution, the Cl^- anion will inhibit the uptake of other anions, particularly nitrate (NO_3^-). Chlorine exists in the nutrient solution as the chloride anion.

Copper

Content in Plants—Sufficiency of copper (Cu) ranges from 2 to 10 mg/kg (ppm) in the dry matter.

Function—Copper plays a role in photosynthesis, as a constituent of a chloroplast protein, and is also known to be an enzyme activator.

Deficiency Symptoms—When deficient, plants are stunted and chlorosis develops on the older leaves. In fruiting crops, Cu deficiency affects the developing fruit; they are small and imperfectly formed. Death of the growing tip of the fruit may also occur with Cu deficiency.

Excess Symptoms—In hydroponic systems, Cu toxicity can result in significant root damage if the Cu content of the nutrient solution is too high [>0.1 mg/L (ppm)].

Content in Nutrient Solution—The normal concentration range for Cu in nutrient solutions ranges from 0.001 to 0.01 mg/L (ppm). Copper exists in the nutrient solution as the cupric (Cu^{2+}) cation. It has been suggested by some that if the Cu concentration is raised to 4 mg/L in nutrient flow systems, some degree of fungus control can be obtained. Additional research is needed to determine if such Cu levels will indeed control common root diseases and not damage plant roots. Such Cu concentrations should not be used for other types of hydroponic growing systems.

Reagent Sources—Copper sulfate ($CuSO_4 \cdot 5H_2O$) is the common reagent source for Cu in nutrient solution formulas. However, there may be sufficient Cu contamination from contact with Cu-containing equipment (pipes, etc.) to supply all that is required in the nutrient solution.

Iron

Content in Plants—The sufficiency range for iron (Fe) in most crops is from 50 to 100 mg/kg (ppm) of the dry matter; the so-called *critical* concentration is 50 mg/kg (ppm). Iron accumulates in plants without any apparent deleterious effect; therefore, it is not unusual to find Fe concentrations in excess of many hundreds of milligrams per kilogram (ppm). Total Fe in the plant may be of little importance, somewhat similar to calcium, in that the soluble or *labile* concentration determines sufficiency. Special tests have been developed to measure this form of Fe in plant tissue (Bar-Akiva et al., 1978; Bar-Akiva, 1984).

Function—Iron plays a significant role in various energy transfer functions in the plant due to ease of valence change ($Fe^{2+} = Fe^{3+} + e^-$). It also has the tendency to form chelate complexes. Iron plays an important role in the process of photosynthesis and the formation of chlorophyll, but other exact roles are not clearly known.

Deficiency Symptoms—One of the symptoms of Fe deficiency is a loss of the plant's green color due to the loss of chlorophyll, the green pigment. Although the appearance of Fe deficiency is not too dissimilar to that of magnesium, an Fe deficiency symptom first appears in the younger plant tissue, whereas magnesium deficiency symptoms first appear in the older tissue. Iron deficiency symptoms are not always clearly distinct and can be easily confused with other elemental deficiencies, as deficiencies of sulfur, manganese, and zinc frequently produce leaf and plant symptoms that are not easily differentiated visually from those of Fe; therefore, the importance of confirming an Fe deficiency by means of a plant analysis or tissue test is underscored.

Iron deficiency, once developed, is very difficult to correct. There is evidence that in some instances Fe deficiency may be genetically controlled, with specific individual plants incapable of normal Fe metabolism and therefore unresponsive to correction. Some plant species, as well as individuals within a species, can respond to Fe-deficient conditions as their roots release hydrogen (H^+) ions to acidify the area immediately surrounding the root and/or release Fe-complexing substances (i.e., siderophores). Plants that are able to modify their immediate root environment have been designated Fe efficient.

Although the use of Fe chelates has markedly improved the control of Fe deficiency, deficiency correction is still a major problem in many crops and growing situations. Iron deficiency may be easier to control hydroponically than in other systems of growing. Soilless culture systems that employ an organic rooting medium are particularly susceptible to Fe deficiency. This difficulty will be discussed in greater detail later.

Content in Nutrient Solution—Most hydroponic formulas call for the use of a chelated form of Fe (such as FeEDTA and FeDTPA) to ensure that its presence in the nutrient solution is as an available form. Iron easily complexes with many substances, which makes Fe concentration difficult to maintain. Since Fe is a common contaminant found nearly everywhere, it may be present in sufficient concentration to prevent deficiency. Normally, the Fe concentration must be maintained at about 2 to 3 mg/L (ppm) in the nutrient solution to prevent deficiency. Iron may exist as either the ferric (Fe^{3+}) or ferrous (Fe^{2+}) cation, depending on the characteristics of the nutrient solution. There may be sufficient Fe in a nutrient solution, depending on the water source and contact with Fe-based piping and other similar materials.

Forms of Utilization—Plants can use either ionic form, although that taken in as ferric-Fe (Fe^{3+}) must be reduced to the ferrous (Fe^{2+}) form. Ferric-Fe can form complexes and precipitates quite easily in the nutrient solution, thereby reducing its concentration and, therefore, availability to plants. It is evident that the chemistry of Fe in the nutrient solution and its uptake by plants are quite complex. In addition, utilization of Fe varies among plant species, as some have the ability to alter the character of the nutrient solution in the immediate vicinity of their roots, thereby influencing Fe availability. Such influences and their effect on plants are discussed in Chapter 4.

Reagent Sources—Although FeEDTA (more recently FeDTPA) is the most common reagent form for use in nutrient solutions, other Fe compounds are also suitable. Iron sulfate ($FeSO_4 \cdot 7H_2O$) and iron phosphate [$Fe_3(PO_4)_8 \cdot 8H_2O$] are the two inorganic forms of Fe, whereas iron citrate and iron tartrate are the two organic types. Of these four compounds, iron citrate is probably the most frequently used, other than FeEDTA, in the more commonly recommended nutrient solution formulas.

As a general rule, it takes less (about half) Fe as FeEDTA to provide the same level of availability as Fe in other non-chelated chemical forms. When using FeEDTA, it is important not to apply more than that recommended because high concentrations of EDTA can be toxic to plants.

Another chelated Fe form is iron lignin sulfonate, which is frequently used in garden-type formulas.

Manganese

Content in Plants—The sufficiency range of manganese (Mn) is from 20 to 100 mg/kg (ppm) of the dry matter for most crops. Plant species sensitive to Mn deficiency are usually equally sensitive to Mn toxicity.

Function—The function of Mn in the plant is not too different from that of iron; it is associated with the oxidation-reduction processes in the photosynthetic electron transport system.

Deficiency Symptoms—Manganese deficiency symptoms first appear on the younger leaves as an interveinal chlorosis, not too dissimilar to symptoms of magnesium deficiency, which first appears on the older leaves. In some instances, plants may be Mn deficient (moderate visual symptoms present) and yet plant growth will be little affected. However, when the deficiency is severe, significant reduction in plant growth can occur. Manganese deficiency can be easily corrected with a foliar application of Mn or by additions of a suitable form of Mn to the rooting media.

Excess Symptoms—Initial Mn excess may produce toxicity symptoms not too dissimilar to deficiency symptoms. With time, toxicity symptoms are characterized by brown spots on the older leaves, sometimes seen as black specks on the stems or fruit. It is not unusual for typical iron deficiency symptoms to appear when Mn is in excess. This similarity can result in improper diagnosis, which can only be resolved by means of a plant analysis.

Content in Nutrient Solution—Hydroponic formulas call for a Mn concentration of 0.5 mg/L (ppm) in the nutrient solution. Since Mn can be easily taken up by plants, care should be exercised to prevent the application of excessive quantities of Mn in the nutrient solution. Manganese exists in the nutrient solution as the manganous (Mn^{2+}) cation, although other oxidation states can be present under varying conditions of oxygen (O_2) supply.

Reagent Sources—The primary reagent source is manganese sulfate ($MnSO_4 \cdot H_2O$).

Molybdenum

Content in Plants—Plant molybdenum (Mo) requirement is very low; the critical level is less than 0.5 mg/kg (ppm) of the dry matter. The Mo concentration found in normally growing plants is usually between 0.5 to 1 mg/kg (ppm), but it may be considerably greater with no apparent toxic effect on the plant itself.

Function—Molybdenum is an essential component of two major enzymes involved in nitrogen metabolism. Nitrogen (N_2) fixation by symbiotic nitrogen-fixing bacteria requires Mo, and the reduction of the nitrate (NO_3^-) ion by the enzyme *nitrate reductase* requires Mo. Therefore, plants receiving all of their nitrogen by root absorption of the ammonium (NH_4^+) cation either do not require Mo or have a reduced requirement for Mo.

Deficiency Symptoms—Molybdenum deficiency symptoms are unique in some ways, sometimes giving the appearance of nitrogen deficiency. Plant growth and flower development are restricted. Cruciferae species are more sensitive to Mo deficiency than other species. Whiptail of cauliflower is probably the most commonly known Mo deficiency.

Content in Nutrient Solution—Hydroponic formulas call for 0.05 mg/L (ppm) Mo in the nutrient solution. Molybdenum exists in the nutrient solution as the molybdate (MoO_4^{2-}) anion.

Reagent Sources—Ammonium molybdate $[(NH_4)_6Mo_7O_{24} \cdot 4H_2O]$ is the common reagent source.

Zinc

Content in Plants—The sufficiency range of zinc (Zn) for most crops is from 15 to 50 mg/kg (ppm) of the dry matter. Zinc is unique in that the critical level in many crops is 15 mg/kg (ppm). At around 15 mg/kg (ppm), a difference of 1 to 2 mg/kg (ppm) can mean the difference between normal and abnormal growth. Precise measurement of the Zn concentration in the plant when doing a plant analysis determination is therefore critical.

Function—Zinc is an enzyme activator, involved in the same enzymatic functions as manganese and magnesium. Only carbonic anhydrase has been found to be specifically activated by Zn. While Zn probably performs additional roles, these other roles are not well understood. Considerable research has been done on the relationships between Zn and phosphorus and between Zn and iron. The results suggest that excessive phosphorus concentrations in the plant interfere with normal Zn function, while high Zn concentrations interfere with iron usage, and possibly vice versa.

Deficiency Symptoms—Zinc deficiency symptoms appear as chlorosis in the interveinal areas of new leaves, producing a banding appearance on some plant leaves. In fruit and nut trees, rosetting occurs at the branch terminals with considerable dieback. Plant and leaf growth become stunted, and when the deficiency is severe, leaves die and fall off. Moderate Zn deficiency symptoms may be confused with symptoms caused by deficiencies of magnesium, iron, or manganese. Therefore, a plant analysis is required to determine which element is deficient.

Excess Symptoms—Many plant species are tolerant to fairly high levels of Zn in their tissues without untoward consequences. These species may contain Zn at concentrations in excess of several hundred parts per million without noticeable detrimental effect. However, for those species that are sensitive to both iron and Zn, such high levels of Zn may induce iron deficiency.

Content in Nutrient Solution—Hydroponic formulas call for Zn at 0.05 mg/L (ppm) in the nutrient solution. Zinc exists in the nutrient solution as the divalent (Zn^{2+}) cation.

Reagent Sources—Zinc sulfate ($ZnSO_4 \cdot 7H_2O$) is the common reagent source.

The essential elements, their form in solution, and common chemical sources for preparing nutrient solutions are listed in Table 7.

Table 7 Essential elements, ionic form in the nutrient solution, and common reagent sources

Essential Element	Ionic Form in the Nutrient Solution	Common Reagent Sources Name	Formula	Element (%)
Major Elements				
Nitrogen (N)	NO_3^-	Ammonium nitrate	NH_4NO_3	N(16) as NO_3
		Calcium nitrate	$Ca(NO_3)_2 \cdot 4H_2O$	N(15) Ca(19)
		Nitric acid	HNO_3	N(15)
		Potassium nitrate	KNO_3	N(13) K(39)
	NH_4^+	Ammonium nitrate	NH_4NO_3	N(16) as NH_4
		Ammonium phosphate (mono)	$NH_4H_2PO_4$	N(11) P(21)
		Ammonium phosphate (di)	$(NH_4)_2HPO_4$	N(18) P(21)
		Ammonium sulfate	$(NH_4)_2SO_4$	N(21) S(24)
Phosphorus (P)	PO_4^{3-}	Ammonium phosphate (mono)	$NH_4H_2PO_4$	P(21) N(11)
		Ammonium phosphate (di)	$(NH_4)_2HPO_4$	P(21) N(18)
		Potassium phosphate (mono)	KH_2PO_4	P(23) K(28)
		Potassium phosphate (di)	K_2HPO_4	P(18) K(45)
		Phosphoric acid	H_3PO_4	P(34)
Potassium (K)	K^+	Potassium chloride	KCl	K(50) Cl(47)
		Potassium nitrate	KNO_3	K(36) N(13)
		Potassium phosphate (mono)	KH_2PO_4	K(28) P(23)
		Potassium phosphate (di)	K_2HPO_4	K(45) P(18)
		Potassium sulfate	K_2SO_4	K(42) S(17)
Calcium (Ca)	Ca^{2+}	Calcium chloride	$CaCl_2$	Ca(36) Cl(64)
		Calcium nitrate	$Ca(NO_3)_2 \cdot 4H_2O$	Ca(19) N(15)
		Calcium sulfate	$CaSO_4 \cdot 2H_2O$	Ca(23) S(19)

Magnesium (Mg)	Mg^{2+}	Magnesium sulfate	$MgSO_4 \cdot 7H_2O$	Mg(10) S(23)
Sulfur (S)	SO_4^{2-}	Ammonium sulfate	$(NH_4)_2SO_4$	S(24) N(21)
		Calcium sulfate	$CaSO_4 \cdot 2H_2O$	S(19) Ca(23)
		Magnesium sulfate	$MgSO_4 \cdot 7H_2O$	S(23) Mg(10)
		Potassium sulfate	K_2SO_4	S(17) K(42)
Micronutrients				
Boron (B)	BO_3^{3-}	Boric acid	H_3BO_3	B(16)
Copper (Cu)	Cu^{2+}	Copper sulfate	$CuSO_4 \cdot 5H_2O$	Cu(25) S(13)
Iron (Fe)	Fe^{3+}, Fe^{2+}	Iron chelate	FeEDTA	Fe(6 to 12)
		Iron citrate	—	Fe(6)
		Iron tartrate	—	
		Ferrous sulfate	$FeSO_4 \cdot 7H_2O$	Fe(20) S(11)
Manganese (Mn)	Mn^{2+}	Manganese sulfate	$MnSO_4 \cdot H_2O$	Mn(24) S(14)
Molybdenum (Mo)	MoO_4^{2-}	Ammonium molybdate	$(NH_4)_6Mo_7O_{24} \cdot 4H_2O$	Mo(8)
Zinc (Zn)	Zn^{2+}	Zinc sulfate	$ZnSO_4 \cdot 7H_2O$	Zn(22) S(11)

The Beneficial Elements   6

The number of elements presently considered essential for the proper nutrition of the higher plants stands at 16; the last element added to the list was chlorine in 1954 (see Table 3). Some plant physiologists feel that the criteria for essentiality established by Arnon and Stout (1939) could preclude the addition of other elements, as these 16 include most of the elements found in substantial quantities in plants. However, there may be some elements that will yet prove essential, as their requirements are at such low levels that it will take considerable sophisticated chemical analysis to uncover them. The question is what elements these would likely be and where the best place to start would be.

It was recognized early that there may be elements that should be included in the nutrient solution which are not considered essential; therefore, the A–Z Micronutrient Solution was developed (see Table 12). Eight of the 20 elements included in the A–Z Micronutrient Solution are considered essential for animals:

arsenic (As)	iodine (I)
cobalt (Co)	nickel (Ni)
chromium (Cr)	selenium (Se)
fluorine (F)	vanadium (V)

Many feel that those elements recognized as essential for animals (see Table 6) but not currently for plants are good candidates for essentiality in plants. Those who may wish to explore the potential for discovery of additional elements that may prove essential for both animals and plants will find the articles by Mertz (1981), Asher (1991), and Pais (1992) interesting.

Four elements are currently being studied for their potential essentiality in plants: cobalt, nickel, silicon, and vanadium. Considerable research has been devoted to each of these elements, and some investigators feel that they are important elements for sound plant growth.

Cobalt

Cobalt (Co) is required indirectly by leguminous plants because this element is essential for the *Rhizobium* bacteria which live symbiotically in the roots, fixing atmospheric nitrogen and providing the host plant with its major source of nitrogen. Without Co, the *Rhizobium* is inactive and the legume plant requires an inorganic source of nitrogen [such as nitrate (NO_3^-) and/or ammonium (NH_4^+) ions] from the soil. It is not clear whether the plant itself requires Co to carry out specific biochemical processes. The irony of this relationship between *Rhizobium* bacteria and leguminous plants is that in the absence of inorganic nitrogen in the soil, which forces the plant to depend wholly on N_2 fixed by the *Rhizobium* bacteria, the plant will be deficient in nitrogen, cease to grow, and eventually die if Co is absent.

Silicon

Silicon (Si) has been found to be required to maintain stalk strength in rice and other small grains (Takahashi et al., 1990). In the absence of adequate Si, these grain plants will not grow upright. This tendency to lodge results in significant grain loss in commercial production situations. The problem of lodging has been observed primarily in paddy rice, where soil conditions may affect Si availability and uptake. Recent studies with greenhouse-grown tomato and cucumber have shown that without adequate Si, plants are less vigorous and unusually susceptible to fungus disease attack (Belanger et al., 1995). Best growth is obtained when the nutrient solution contains 100 mg/L (ppm) of silicic acid (H_2SiO_3).

Nickel

Nickel (Ni) is considered an essential element for both legumes and small grains (i.e., barley) because Brown et al. (1987) have shown that its deficiency meets the requirements for essentiality established by Arnon and Stout (1939). Nickel is a component of the enzyme urease, and plants deficient in Ni have high accumulations of urea in their leaves. Nickel-deficient plants are slow growing, and for barley, viable grain is not produced. It is recommended that a nutrient solution contain a Ni concentration of at least 0.057 mg/L (ppm) in order to satisfy the plant requirement for this element.

Vanadium

Vanadium (V) seems to be capable of substituting for molybdenum in the nitrogen metabolism of plants, with no independent role clearly established for V.

Other Elements

There is considerable evidence that some non-essential elements can partially substitute for an essential element, such as sodium for potassium and vanadium for molybdenum. These partial substitutions may be quite beneficial to plants in situations where an essential element is at a marginally sufficient concentration. For some plant species, this partial substitution seems to be beneficial to the plant. Despite considerable speculation, it is not known exactly how and why such substitutions take place.

There seem to be elements that are beneficial to plants, but their exact function and optimum concentration level are as yet unknown (Asher, 1991; Pais, 1992). This is one of the justifications for the A–Z Micronutrient Solution (see Table 12) as a supplement to the nutrient solution. This situation presents real problems for plant physiologists as well as growers when using nutrient solutions as the source of supply for the essential elements. The absence or low level of one or more of these non-essential elements in the nutrient solution could have marked effects on plant growth and development. There is gathering evidence that a number of elements beyond the four described above may be beneficial to plants at low levels but toxic at higher concentrations. Even elements considered very toxic to plants, such as lead and platinum (Pallas and Jones, 1978), have been found to have a stimulating effect on plant growth at very low concentrations—between 10 and 100 µg/L [parts per billion (ppb)].

The growing evidence of beneficial effects from elements that are not currently recognized as essential for plants should be sufficient to alert growers that the use of pure reagents and purified water for making the nutrient solution may not be the best practice. The presence of small quantities of elemental impurities may be desirable. Consideration should, therefore, be given to including them in the nutrient solution. The use of the A–Z Micronutrient Solution (see Table 12) is again encouraged.

Humic Acid

Humic acid and its potential role in plant nutrition have gained considerable interest in recent years. Humic acid is extracted from soil organic matter or peat, and it is a material with interesting physiochemical properties that have been found to enhance plant growth (Tan, 1993). The addition of humic acid to the nutrient solution has been proposed, but its benefits have yet to be thoroughly evaluated. It is believed that humic acid would chelate some elements in the nutrient solution, thereby providing some control in elemental uptake and utilization.

The Nutrient Solution

7

Probably no aspect of hydroponic/soilless growing is as poorly understood as the constitution and use of nutrient solutions. Most texts simply provide the reader with a list of nutrient solution formulas, preferred reagent sources, and the necessary techniques for calculating weights and measures. Although such information is surely essential to properly prepare the nutrient solution, a soundly based understanding of the management of the solution is as important, if not more so, for successful growing. The complex interrelationships between composition and use are not understood by many growers, and it is this aspect of nutrient solution management for which much of the literature unfortunately provides little or no help. Poor yields, scraggly plants, high water and reagent costs, indeed most of the hallmarks of a less than fully successful growing operation can be directly linked to mismanagement of the nutrient solution. There are, unfortunately, no absolute pat prescriptions or recipes that can be given to growers by any writer. Growers will have to experiment with their own systems—observing, testing, and adjusting until the proper balance between composition and use is achieved in their particular situation and for specific crops.

In addition, the usual management considerations relating to costs for reagents and water, as well as the energy required to move the nutrient solution, must be integrated into the successful operation of a hydroponic/soilless growing system. One of the major financial decisions involves balancing replenishment schedules against input costs and losses due to single use and dumping.

Although much is unknown about how best to manage the nutrient solution, there are many good clues as to what should or should not be done. This chapter is devoted to an explanation of these clues. Growers using these clues will have to develop a scheme of management which best fits their environmental system and crops. They will have to experiment with various techniques to obtain maximum utilization of the nutrient solution while achieving high crop yields of top quality.

Terms used to describe the two methods of nutrient solution management are open and closed. An *open system* is one in which the nutrient solution is used only once in a one-way passage through the rooting vessel. In a *closed system,* the nutrient solution is reused by recirculation (Hurd et al., 1980). These two means of nutrient solution management pose different problems for the grower, which will be discussed in greater detail later.

Water Quality

In many parts of the United States and indeed throughout the world, water quality can be a major problem for hydroponic/soilless culture use due to contamination by various inorganic and organic substances. Even water supplies suitable for domestic and/or agricultural use must be considered suspect by the hydroponic/soilless grower. Since most hydroponic/soilless culture systems require sizable quantities of relatively pure water, water quality is of utmost importance. The best domestic water supplies commonly contain substances and elements that can affect plant growth. Even rainwater collected from the greenhouse covering may contain substances that can affect plant growth.

Therefore, a complete analysis of the water to be used for any type of hydroponic/soilless culture system is essential. The analysis should include both organic as well as inorganic components if the water is being taken from a river, shallow well, or other surface source. When taken from sources other than these, an inorganic elemental assay will be sufficient to determine elemental composition and concentration.

Natural water supplies can contain sizable concentrations of some of the essential elements required by plants, particularly calcium and magnesium. In areas where water is being taken from limestone-based aquifers, it is not unusual for concentrations of calcium and magnesium to be as high as 100 and 30 mg/L (ppm), respectively. Some natural waters will contain sizable concentrations of sodium and anions such as bicarbonate (HCO_3^-), carbon-

Table 8   Maximum mineral concentrations for irrigation water used in rockwool culture

Element/Ion	*Maximum Concentration (mg/L, ppm)*
Chloride (Cl)	50–100
Sodium (Na)	30–50
Carbonate (CO_3)	4.0
Boron (B)	0.7
Iron (Fe)	1.0
Manganese (Mn)	1.0
Zinc (Zn)	1.0

Source: Verwer and Wellman, 1980.

ate (CO_3^{2-}), sulfate (SO_4^{2-}), and chloride (Cl^-). In some areas, boron may be found in fairly high concentrations. Sulfide (S^-), primarily as iron sulfide, which gives the "rotten egg" smell to water, is found in some natural waters.

Verwer and Wellman (1980) have established what the maximum mineral concentration would be for water used in rockwool culture, as shown in Table 8. In addition, Waters et al. (1972) have set the suitability of water for irrigating pot plants, and Farnhand et al. (1985) have established criteria for irrigation water based on salinity, ion content, and effect on crops; their data are given in Tables 9 and 10.

Surface or pond water may contain disease organisms or algae which pose problems. Algae grows extraordinarily well in most hydroponic cul-

Table 9   Suitability of water for irrigating potted plants

Water Classification	*Electric Conductance (mmhos/cm)*	*Total Dissolved Solids (salts) (mg/L, ppm)*	*Sodium (% of total solids)*	*Boron (mg/L, ppm)*
Excellent	<0.25	<175	<20	<0.33
Good	0.25–0.75	175–525	20–40	0.33–0.67
Permissible	0.75–2.0	525–1400	40–60	0.67–1.00
Doubtful	2.0–3.0	1400–2100	60–80	1.00–1.25
Unsuitable	>3.0	>2100	>80	>1.25

Source: Waters et al., 1972.

Table 10   Water quality guidelines for irrigation

Characteristic	Degree of Problem		
	None	Increasing	Severe
EC, dS/m*	< 0.75	0.75–3.0	>3.0
TDS, mg/L**	<480	480–1920	>1920
Sodium (Na), SAR value	<3	3–9	>9
Chloride (Cl), mg/L	<70	70–345	>345
Boron (B), mg/L	1.0	1.0–2.0	2.0–10.0
Ammonium (NH_4) and nitrate (NO_3), mg/L	<5	5–30	>30
Bicarbonate (HCO_3), mg/L	<40	40–520	>520

* Electrical conductance.
** Total dissolved solids.

Source: Farnhand et al., 1985.

ture systems, plugging pipes and fouling valves. Filtering and/or other forms of pretreatment are required to ensure that the water used to prepare the nutrient solution is free from these undesirable organisms and suspended matter.

In most cases, some form of water treatment will be necessary to make and maintain useful solutions. Depending on what the analysis of the water supply indicates, the grower may at one end of the range simply have to filter out debris; at the other extreme, sophisticated systems dedicated to ion removal by means of ion exchange or reverse osmosis as well as filtering using sand beds or fine-pore-type filters may be required (Anon., 1997).

In hard-water areas, there may be sufficient calcium and magnesium in the water to provide a portion or all of the plant requirements, or the micronutrient element concentration could be sufficient to preclude the need to add this group of elements to the nutrient solution. These determinations can and should be made only on the basis of an elemental analysis of the water.

The presence of organic chemicals, such as pesticides and herbicides, can significantly affect plant growth if present in sufficient quantity in water. Water from shallow wells, or from surface water sources in intensively cropped agricultural areas, should be tested for the presence of these types of chemicals.

Treatment should be employed only if the chemical and/or physical composition of the water warrants. Obviously, business, financial, and managerial planning must incorporate the costs of developing nutrient pure water in a grower's specified environment. For example, it may be financially prudent to accept some crop loss from the use of impure water rather than attempting to recover the cost of water treatment; treatment may be as simple and inexpensive a task as acidifying the water to remove bicarbonates (HCO_3) and carbonates (CO_3).

Water samples should be submitted to a testing laboratory for a complete analysis before use, and the analysis should be repeated whenever a change in the water source is made. It is also advisable to have the initial nutrient solution assayed to be sure that its composition is as expected before its use. Instrumental devices and analysis kits can be used when monitoring water and nutrient solutions. For example, pH and conductivity meters are handy for self-monitoring of the nutrient solution. The HACH Chemical Company (P.O. Box 389, Loveland, CO 80539) has one of the best kits available for the grower to use in monitoring the composition of water and nutrient solution. When using any testing device or kit, the user must carefully follow the instructions provided, and it is advisable to have reference samples of known composition on hand for verification of the accuracy of the determinations made on water and nutrient solution.

Water and Nutrient Solution Filtering

Any suspended material in the water source should be removed by filtering the water through a sand bed or similar filter system. Suspended material may carry disease organisms, be a source for algae, or precipitate some elements in reagents when constituting the nutrient solution.

With continuous use in a closed recirculating system, the nutrient solution is altered, not only chemically through the removal of elements by plant root absorption but also through additions produced by the decay of plant material, the reverse action of plant roots, and the release or development of substances contained in or incident to the support media. As a result, the nutrient solution may become cloudy as suspended precipitates, microorganisms, and algae are added to it as it passes through the rooting medium.

For short-term use (less than seven days), a change in the appearance of the nutrient solution is of little consequence. However, if the nutrient solution is to be used for an extended period of time (>5 days) and additions of spent elements are made to extend its use, then these changes require special attention. Filtering is the way in which this problem is best resolved.

The grower has a number of options in filtering the nutrient solution. Swimming-pool-type filtering systems are capable of removing suspended particles of 50 μm and larger. Removal of particles below 50 μm requires the installation of a sophisticated filtering system, such as Millipore filters (Millipore Corporation, Ashby Road, Bedford, MA 01730). Such a system is capable of removing substances that are microscopic in size (less than 1 μm). Thus, such a system removes not only large contaminants but also a number of disease organisms from the nutrient solution.

Filtering the nutrient solution is not a common practice, nor is it recommended in most of the literature on hydroponics. The only exception would be water dispensed through a drip system, which must be free of suspended substances or filtered. Most nutrient solution management schemes simply call for dumping the *spent* nutrient solution frequently. If water supplies are limited, nutrient solution reuse may be necessary, making filtering an important procedure to maintain the solution in usable form.

Unfortunately, sophisticated filtering systems are expensive and require close attention to keep in proper operating condition. They also add to the cost of growing plants hydroponically. Therefore, what may be gained by filtering must be evaluated against its added cost. Also, as there is little research or practical information available to adequately evaluate cost versus improvement in plant performance, the grower must make the analysis in terms of conditions and conservative assumptions.

Size, type, and installation requirements for a filtering system will vary depending on water volume, frequency of use, and quantity of material accumulating in the nutrient solution. Cartridge-type filters are recommended, as back-flushing is not generally possible or practical with most hydroponic systems, and cartridges can be easily removed and exchanged. Filtering devices should be placed in the outflow pipe leading to the growing bed from the supply reservoir or container. The coarser filter should be placed first in line, followed by the finer one.

To provide some degree of control over microorganisms (bacteria, etc.), in addition to the use of a Millipore filter, the nutrient solution can be passed under ultraviolet radiation (Buyanovsky et al., 1981; Evans, 1995). UV sterilizers have proven to be effective in reducing microorganism counts when two 16-watt lamps are placed in the path of the nutrient solution flowing at 13.5 L (3 gallons) per minute, giving a total exposure of 573 joules per square meter per hour.

Nutrient Solution Formulas

Probably no single aspect of soilless culture is as poorly understood as the constitution of the nutrient solution and its management. While it is true that numerous formulas for preparing the nutrient solution have been published, their proper use relative to the growing system and needs of a specific plant species has been largely ignored. The formulas devised by Hoagland and Arnon (Table 11) are widely used—generally in modified form. It is common to see the phrase "modified Hoagland's nutrient solution" in the literature; it refers to their frequently cited University of California Circular 347 (Hoagland and Arnon, 1950). The only difference be-

Table 11   Hoagland's nutrient solution formulas

Stock Solution	To use: mL/L
Solution No. 1	
$1M$ potassium dihydrogen phosphate (KH_2PO_4)	1.0
$1M$ potassium nitrate (KNO_3)	5.0
$1M$ calcium nitrate [$Ca(NO_3)_2 \cdot 4H_2O$]	5.0
$1M$ magnesium sulfate ($MgSO_4 \cdot 7H_2O$)	2.0
Solution No. 2	
$1M$ ammonium dihydrogen phosphate ($NH_4H_2PO_4$)	1.0
$1M$ potassium nitrate (KNO_3)	6.0
$1M$ calcium nitrate [$Ca(NO_3)_2 \cdot 4H_2O$]	4.0
$1M$ magnesium sulfate ($MgSO_4 \cdot 7H_2O$)	2.0
Micronutrient Stock Solution	*g/L*
Boric acid (H_3BO_3)	2.86
Manganese chloride ($MnCl_2 \cdot 4H_2O$)	1.81
Zinc sulfate ($ZnSO_4 \cdot 5H_2O$)	0.22
Copper sulfate ($CuSO_4 \cdot 5H_2O$)	0.08
Molybdate acid ($H_2MoO_4 \cdot H_2O$)	0.02
	To use:
	1 mL/L nutrient solution
Iron	
For Solution No. 1: 0.5% iron ammonium citrate	to use: 1 mL/L
For Solution No. 2: 0.5% iron chelate	to use: 2 mL/L

Source: Hoagland and Arnon, 1950.

tween Hoagland Solutions 1 and 2 is that Solution No. 2 contains a portion of its nitrogen content as ammonium (NH_4), while in Solution No. 1, all of the nitrogen is in the nitrate (NO_3) form. A number of nutrient solution formulas in the earlier literature on hydroponics are given in Table 12.

Table 12   List of formulas for constituting nutrient solutions from the early literature on hydroponics

Name	Reagents	g/L (mg/mL)		
Knop's Solution[1]	KNO_3	0.2		
	$Ca(NO_3)_2$	0.8		
	KH_2PO_4	0.2		
	$MgSO_4 \cdot 7H_2O$	0.2		
	$FePO_4$	0.1		
		a	b	c
Crone's Solution (1902, 1904)[1]	KNO_3	1.00	0.75	0.75
	$Ca_3(PO_4)_2$	0.25	0.25	0.25
	$CaSO_4 \cdot 2H_2O$	0.25	0.25	0.50
	$Fe_3(PO_4)_2 \cdot 8H_2O$	0.25	0.25	0.25
	$MgSO_4 \cdot 7H_2O$	0.25	0.25	0.50
Hoagland and Snyder (1933)[1]	KNO_3	0.31		
	$Ca(NO_3)_2$	0.82		
	$MgSO_4 \cdot 7H_2O$	0.49		
	KH_2PO_4	0.136		
	Ferric tartrate, 1 mL/L of 0.5% solution			
	Micronutrients: A–Z Solution			
Trelease and Trelease (1933)[1]	KNO_3	0.683		
	$(NH_4)_2SO_4$	0.0679		
	KH_2PO_4	0.3468		
	K_2HPO_4	0.01233		
	$CaCl_2$	0.4373		
	$MgSO_4 \cdot 7H_2O$	0.7478		
	$FeSO_4 \cdot 7H_2O$	0.00278		
		a	b	c	Modified Crone's
The Original Rothamsted Solutions[1]	KNO_3	1.0	1.0	1.0	1.0
	$MgSO_4 \cdot 7H_2O$	0.3	0.3	0.5	0.5
	KH_2PO_4	0.45	0.4	0.3	—
	K_2HPO_4	0.0675	0.133	0.27	—

Table 12   List of formulas for constituting nutrient solutions from the early literature on hydroponics (continued)

Name	Reagents	g/L (mg/mL)			
	$CaSO_4 \cdot 2H_2O$	0.5	0.5	0.5	0.5
	$Ca_3(PO_4)_2$	—	—	—	0.25
	$Fe_3(PO_4)_2 \cdot 8H_2O$	—	—	—	0.25
	$FeCl_3$	0.04	0.04	0.04	—
	H_3BO_3	0.001	0.001	0.001	0.001
	$MnSO_4 \cdot 4H_2O$	0.001	0.001	0.001	0.001
Amon (1938)[1]	KNO_3	0.656			
	$Ca(NO_3)_2$	0.656			
	$NH_4H_2PO_4$	0.115			
	$MgSO_4 \cdot 7H_2O$	0.49			
	$FeSO_4 \cdot 7H_2O$	0.5%			
	Tartaric acid—0.4%	0.6 mL/1.3× weekly			
	H_3BO_3	2.86 mg			
	$MnCl_2 \cdot 4H_2O$	1.81 mg			
	$CuSO_4 \cdot 5H_2O$	0.08 mg			
	$ZnSO_4 \cdot 7H_2O$	0.22 mg			
	$H_2MoO_4(MoO_3 + H_2O)$	0.09 mg			
Arnon and Hoagland (1940)[1]	KNO_3	1.02			
	$Ca(NO_3)_2$	0.492			
	$NH_4H_2PO_4$	0.230			
	$MgSO_4 \cdot 7H_2O$	0.49			
	See: Arnon's Micronutrient Formula				
Shive and Robbins (1942) I[1]	$Ca(NO_3)_2$	0.938			
	$(NH_4)_2SO_4$	0.0924			
	KH_2PO_4	0.313			
	$MgSO_4 \cdot 7H_2O$	0.567			
	$FeSO_4 \cdot 7H_2O$	5.50 mg			
	H_3BO_3	0.57 mg			
	$MnSO_4 \cdot 4H_2O$	0.57 mg			
	$ZnSO_4 \cdot 7H_2O$	0.57 mg			
Shive and Robbins (1942) II[1]	$NaNO_3$	0.34			
	$CaCl_2$	0.1665			
	KH_2PO_4	0.214			
	$MgSO_4 \cdot 7H_2O$	0.514			
	Iron and micronutrients as in I				
Piper (1942)[1]	KNO_3	1.3			
	KH_2PO_4	0.3			
	NaCl	0.1			

Table 12 List of formulas for constituting nutrient solutions from the early literature on hydroponics (continued)

Name	Reagents	g/L (mg/mL)
	$CaSO_4 \cdot 2H_2O$	0.3
	$MgSO_4 \cdot 7H_2O$	0.5
	Ferric citrate	0.02
	H_3BO_3	0.3 mg
	Mn (as $MnSO_4$)	0.5 mg
	Zn (as $ZnSO_4$)	0.2 mg
	Mo (as Na_2MoO_4)	0.1 mg
	Cu (as $CuSO_4$)	0.003 mg
Robbins (1946)[1]	KNO_3	0.408
	$Ca(NO_3)_2$	0.820
	KH_2PO_4	0.136
	$MgSO_4 \cdot 7H_2O$	0.493
	Fe	0.30 mg
	B	0.25 mg
	Mn	0.25 mg
	Zn	0.25 mg
	Cu	0.02 mg
	Mo	0.01 mg
Kuwait IV[2]	$MgSO_4 \cdot 7H_2O$	0.34
	$CaHPO_4$	0.13
	$Ca(NO_3)_2$	2.0
	KNO_3	0.26
	K_2SO_4	0.02
	NaCl	0.15
	HNO_3 (conc)	13.00 mL
	HCl (conc)	20.00 mL
Kuwait IV (D.W.2)[2]	$MgSO_4 \cdot 7H_2O$	0.34
	KH_2PO_4	0.13
	$Ca(NO_3)_2$	2.09
	KNO_3	0.16
	K_2SO_4	0.02
	NaCl	0.15
	HNO_3 (conc)	13.00 mL
	HCl (conc)	20.00 mL
Gravel Culture, Japan[2]	KNO_3	0.81
	$Ca(NO_3)_2$	0.95
	$MgSO_4 \cdot 7H_2O$	0.50
	$NH_4H_2PO_4$	0.12

Table 12 List of formulas for constituting nutrient solutions from the early literature on hydroponics continued)

Name	Reagents	g/L (mg/mL)
Basic Formula, Bengal, India[2]	$NaNO_3$	0.17
	$(NH_4)_2SO_4$	0.08
	$CaSO_4$	0.04
	$CaHPO_4$	0.10
	K_2SO_4	0.11
	$MgSO_4 \cdot 7H_2O$	0.07
Rivoira's Formula, Sassari, Sicily[2]	$(NH_4)_2HPO_4$	0.20
	$Ca(NO_3)_2$	0.50
	KNO_3	0.20
	$MgSO_4 \cdot 7H_2O$	0.10
	FeEDTA	5.13 mg
	$MnSO_4 \cdot H_2O$	0.73 mg
	$ZnSO_4 \cdot 7H_2O$	0.06 mg
	$CuSO_4 \cdot 5H_2O$	0.06 mg
	H_3BO_3	0.59 mg
Wroclaw Formula, Poland[2]	KNO_3	0.6
	$Ca(NO_3)_2$	0.7
	NH_4NO_3	0.1
	$CaHPO_4$	0.5
	$MgSO_4 \cdot 7H_2O$	0.25
	$Fe_2(SO_4)_3$	0.12
	H_3BO_3	0.60 mg
	$MnSO_4 \cdot H_2O$	0.60 mg
	$ZnSO_4 \cdot 7H_2O$	0.06 mg
	$CuSO_4 \cdot 5H_2O$	0.30 mg
	$(NH_4)_6Mo_7O_{24} \cdot 4H_2O$	0.06 mg
Volcani Institute, Israel[2]	KNO_3	0.45
	NH_4NO_3	0.35
	$MgSO_4 \cdot 5H_2O$	0.05
	H_3PO_4	100 mL
Penningsfleld's North African Formula[2]	KNO_3	0.38
	$Ca(NO_3)_2$	0.21
	$NH_4H_2PO_4$	0.04
	KH_2PO_4	0.14
	$MgSO_4 \cdot 7H_2O$	0.19
	$Fe_2(SO_4)_3$	0.01
	$Na_2B_4O_7 \cdot 10H_2O$	2.5 mg
	$MnSO_4 \cdot H_2O$	2.5 mg

Table 12   List of formulas for constituting nutrient solutions from the early literature on hydroponics (continued)

Name	Reagents	g/L (mg/mL)
	$CuSO_4 \cdot 5H_2O$	2.5 mg
	$(NH_4)_6Mo_7O_{24} \cdot 4H_2O$	0.75 mg
	$ZnSO_4 \cdot 7H_2O$	0.02 mg
USDA Formula[2]	KNO_3	0.52
Maryland*	$(NH_4)_2SO_4$	0.088
	$CaSO_4$	0.22
	$MgSO_4 \cdot 7H_2O$	0.40
	$CaSO_4$	0.43
Ag. Extension Service[2]	KNO_3	0.36
Florida*	$(NH_4)_2SO_4$	0.08
	$CaHPO_4$	0.17
	$MgSO_4 \cdot 7H_2O$	0.16
	$CaSO_4$	0.90
*Micronutrient Formula[2]	$Fe_2(SO_4)_3$	9.50 mg
for USDA and Ag.	$MnSO_4 \cdot H_2O$	0.63 mg
Extension Service	$CuSO_4 \cdot 5H_2O$	0.29 mg
	$Na_2B_4O_7 \cdot 10H_2O$	7.20 mg
	$ZnSO_4 \cdot 7H_2O$	0.29 mg
Micronutrients A–Z[1]	A. $Al_2(SO_4)_8$	0.055
	KI	0.027
	KBr	0.027
	TiO_2	0.055
	$SnCl_2 \cdot 2H_2O$	0.027
	LiCl	0.027
	$MnCl_2 \cdot 4H_2O$	0.38
	H_3BO_3	0.61
	$ZnSO_4 \cdot 7H_2O$	0.055
	$CuSO_4 \cdot 3H_2O$	0.055
	$NiSO_4 \cdot 6H_2O$	0.055
	$Co(NO_3)_2 \cdot 6H_2O$	0.055
	B. As_2O_3	0.0055
	$BaCl_2$	0.027
	$CdCl_2$	0.0055
	$Bi(NO_3)_2$	0.0055
	Rb_2SO_4	0.0055
	K_2CrO_4	0.027
	KF	0.0035
	$PbCl_2$	0.0055

Table 12   List of formulas for constituting nutrient solutions from the early literature on hydroponics (continued)

Name	Reagents	g/L (mg/mL)
	$HgCl_2$	0.0055
	MoO_3	0.023
	H_2SeO_4	0.0055
	$SrSO_4$	0.027
	VCl_3	0.0055
Arnon's Micronutrient Formula[1]	H_3BO_3	0.48 mg/L
	$MnSO_4 \cdot H_2O$	0.25 mg/L
	$ZnSO_4 \cdot 7H_2O$	0.035 mg/L
	$CuSO_4 \cdot 3H_2O$	0.008 mg/L
	$MoO_3 \cdot 2H_2O$	0.1104 mg/L

Sources: 1. Hewitt, E.J. 1966. *Sand and Water Culture Methods Used in the Study of Plant Nutrition.* Technical Communication No. 22, Revised. Commonwealth Agricultural Bureaux, Maidstone, Kent, England. 2. Douglas, J.S. 1976. *Advanced Guide to Hydroponics.* Drake Publishers, New York.

Lorenz and Maynard (1988) published four researchers' nutrient solution formulas, identifying their use only for commercial greenhouse vegetable production (Table 13).

Elemental Content in the Nutrient Solution

Although the nutrient solution formula may be modified to suit particular requirements for its use, the critical requirements for proper management are either overlooked or not fully understood. The hydroponic literature is marked by much comment on nutrient solution composition in terms of the concentration of the elements in solution but is nearly devoid of instructions as to how the nutrient solution is to be used in such common management terms as the volume per plant and frequency of renewal and replenishment of *spent* elements prior to renewal.

When discussing these questions, Cooper (1988), developer of the nutrient film technique system (Cooper, 1979), remarked that "there is very little information available on this subject." In an interesting experiment, he obtained maximum tomato plant growth when tomato plants were exposed to 60 L (13.3 gallons) of nutrient solution per plant per week. Thinking that

Table 13 Some nutrient solution formulas for commercial greenhouse vegetable production

Reagent	Amount (g/100 gallons of water)			
	Johnson	Jensen	Larson	Cooper
Potassium nitrate	95	77	67	221
Monopotassium phosphate	54	103	—	99
Potassium magnesium sulfate	—	—	167	—
Potassium sulfate	—	—	130	—
Calcium nitrate	173	189	360	380
Magnesium sulfate	95	187	—	194
Phosphoric acid (75%)	—	—	40 mL	—
Chelated iron (FeDTPA)	9	9.6	12	30
Boric acid	0.5	1.0	2.2	0.6
Copper sulfate	0.01	—	0.5	0.15
Copper chloride	—	0.05	—	—
Manganese sulfate	0.3	0.9	1.5	2.3
Zinc sulfate	0.04	0.15	0.5	0.17
Molybdic acid	0.005	0.02	0.04	—
Ammonium molybdate	—	—	—	0.14
Element	Concentration in Solution (mg/L, ppm)			
Major elements				
Nitrogen (N)	105	106	172	236
Phosphorus (P)	33	62	41	60
Potassium (K)	138	156	300	300
Calcium (Ca)	85	93	180	185
Magnesium (Mg)	25	48	48	50
Sulfur (S)	33	64	158	68
Micronutrients				
Boron (B)	0.23	0.46	1.0	0.3
Copper (Cu)	0.01	0.05	0.3	0.1
Iron (Fe)	2.3	3.8	3.0	12.0
Manganese (Mn)	0.26	0.81	1.3	2.0
Molybdenum (Mo)	0.007	0.03	0.07	0.2
Zinc (Zn)	0.024	0.09	0.3	0.1

Source: Lorenz and Maynard, 1988.

growth was enhanced by the removal of root exudate due to the large volume of solution available to the plants, he studied the relationship between root container size and nutrient solution flow rate. He found that plant growth was affected principally by the size of the rooting container and the volume of nutrient solution flowing through the container, not by the removal of root exudates. Cooper concluded that more fundamental research was needed to determine the best volume of nutrient solution and flow characteristics for maximum plant growth. He also observed that "the tolerance of nutrient supply was found to be very great."

This observation seems to be in agreement with Steiner (1980), developer of the Steiner formulas (Table 14), who feels that plants have the ability "to select the ions in the mutual ratio favourable for their growth and development," if they are cultivated in an abundant nutrient flow. Available evidence suggests that an advantage of flowing nutrient solution systems arises from the larger volume of nutrient solution available to the plant, resulting in increased contact with the essential elements and reduction in the concentration of inhibiting substances.

Steiner (1961) has suggested that only a handful of nutrient solution formulas are useful; at best, only one formula is sufficient for most plants as long as the ion balance between the elements is maintained. Steiner feels that most plants will grow extremely well in one *universal* nutrient solution with the following percentage equivalent ratios of anions and cations:

NO_3^-	50 to 70% of the anions
$H_2PO_4^-$	3 to 20% of the anions
SO_4^{2-}	25 to 40% of the anions
K^+	30 to 40% of the cations
Ca^{2+}	35 to 55% of the cations
Mg^{2+}	15 to 30% of the cations

He also suggests that these ion concentration ratios may vary a bit as follows:

NO_3^-	:	$H_2PO_4^-$	:	SO_4^{2-}
60	:	5	:	35
K^+	:	Ca^{2+}	:	Mg^{2+}
35	:	45	:	20

These ion ratios are graphically presented in Figure 3.

Steiner's (1980) thesis depends upon the assumption that plants can adjust to ratios of cations and anions which are not typical of their normal uptake characteristics, but that plants will expend much less energy if the ions of the essential elements are in proper balance as given above. Steiner's thesis explains, in part, why many growers have successfully grown plants using Hoagland-type nutrient solution formulas, as plants are apparently

Table 14 Steiner's universal method for preparing nutrient solutions

Concept: Steiner raises the question as to whether it is the relative concentration of elements among each other or the absolute amount that determines uptake. He suggests that there must be a minimum concentration below which uptake is no longer possible and above which luxury consumption occurs, leading to internal toxicity. However, within this range, there also must be relative relationships that determine uptake and, therefore, the composition of the nutrient solution must be in a particular balance to satisfy the plant requirement for essential elements.

Method: Steiner's aim was to determine how a particular nutrient solution could be prepared which satisfies given requirements as to:

1. relative cation ratios
2. relative anion ratios
3. total ionic concentration
4. pH

Concerning himself with three major anions and three major cations, Steiner established the equivalent ratio in percent as follows:

NO_3^-	:	$H_2PO_4^-$	:	SO_4^{2-}
80		5		15

K^+	:	Ca^{2+}	:	Mg^{2+}
80		10		10

Using five different source reagents and aliquots to establish the desired ratio of ions and not to exceed a total of 30 mg ions per liter, Steiner's formula was:

Reagent	Normality (N)	mL/10 L
KH_2PO_4	1	8.22
$Ca(NO_3)_2 \cdot 4H_2O$	0	1.644
$MgSO_4 \cdot 7H_2O$	2	8.22
KNO_3	1	115.07
K_2SO_4	1	8.22

Preparing various nutrient solutions with the objective of maintaining the desired ratio of ions and pH, it became evident to Steiner that if the total ionic concentration was raised above 30 mg ions per liter and the pH above 6.5, only a very few combinations were possible in order to avoid problems due to precipitation.

Result: Steiner's study reveals the possibility of preparing nutrient solutions that have specific ratios of ions to each other, a set total ion concentration and pH. He sets forth an interesting approach to nutrient solution formula development that bears further study for the inclusion of the other essential elemental forms, such as NO_3-N and NH_4-N, and techniques for use.

Source: Steiner, 1980.

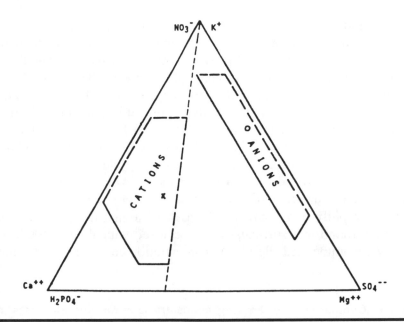

Figure 3   The composition of Steiner's universal nutrient solution (O and X) and the favorable areas for plant production. Solid lines = precipitation limits at 0.7 atm osmotic pressure and pH 6.5; broken lines = physiological limits. (Source: Steiner, 1980.)

able to adjust to the composition of the nutrient solution even when the ratios of ions are not within the ranges required for best growth. Steiner also suggests that the proper balance and utilization of ions in the nutrient solution are best achieved by using his Universal Nutrient Solution formulas (Steiner, 1984). In contrast to the Steiner concept, Schon (1992) has more recently discussed the need to tailor the nutrient solution to meet the demands of the plant.

The Hoagland and Arnon (1950) contributions provide another example of an imperfectly understood and improperly used system. The source of information for both of their nutrient solution formulas was obtained from the determination of the average elemental content of a tomato plant. They calculated the elemental concentration required based on one plant growing in 18 L (4 gallons) of nutrient solution which was replaced weekly. Naturally, one might ask how these nutrient solution formulas would work if tomato is not the crop, the ratio of plant to volume of nutrient solution is changed, and the replenishment schedule is shorter or longer.

Surprisingly, some modification in the use of Hoagland and Arnon formulas seems to have little effect on performance and explains why they are so widely accepted and used. However, significant departure from the plant to nutrient solution volume ratio and replenishment schedule results in either a deficiency or an excess. It is now apparent that if the scheduling of nutrient solution replacement and/or the volume of solution ratio is changed, the nutrient solution formula must be modified to be in accord with the revised practice (i.e., more dilute if the volume per plant ratio and/or frequency of change is increased). This was clearly illustrated by Asher and Edwards (1978a), who devised a hydroponic growing system for carefully controlled plant nutrition studies. In their growing system, the nutrient solution was rapidly and continuously passed through the suspended plant roots. They found that sufficiency was achieved with about 1/100th of the normally recommended elemental concentration in the nutrient solution,

Table 15   Comparisons of limiting concentrations for nine elements in some nutrient solutions commonly used for experimental purposes

| | Parts per Million | | | |
Element	Deficient	Just Adequate	Toxic	Common Range in Nutrient Solutions
Nitrogen (N)				
as nitrate (NO_3)	0.14–10	3.0–70	20–200	49–210
as ammonium (NH_4)	0.007–5	0.03–25	0.4–100	0–154
Potassium (K)				
ammonium present	0.4–6	10–39	—	59–300
ammonium absent	0.04–4	1.1–5	—	
Calcium (Ca)	0.02–22	0.24–40	—	80–200
Magnesium (Mg)	0.05–6	0.2–9	—	24–60
Phosphorus (P)	0.003–4	0.007–2.6	0.03–4	15–192
Sulfur (S)	—	1.3	—	48–224
	Parts per Billion (1/1000 ppm)			
Manganese (Mn)	0.55–71	0.55–2310	16.5–3850	110–550
Zinc (Zn)	0.65–3	3.25–16	195–390	0–146
Copper (Cu)	0.63	1.26	—	0–10

Source: Asher and Edwards, 1978a.

while toxicity occurred if at or near what is normally recommended (Table 15). The range in concentration reflects that required for different plant species used in their studies.

If the primary design of a nutrient solution formula is to provide a certain elemental concentration level in the nutrient solution, what are these levels? For example, the Hoagland-type formula will provide a range in elemental content as shown in Table 16.

Wilcox (1983) determined the elemental composition in a nutrient solution based on various formulas beginning with that of Sacks in 1860, four other formulas, and for two Hydro-Flo formulas (Table 17). As can be seen, there is a fairly wide range in elemental concentration in solution among these formulas. More recently, a similar comparison was made by Gerber (1985) based on more recently recommended formulas, as shown in Table 18; as can be seen, the range in the elemental concentration among the constituted nutrient solutions is still quite large.

Table 16   Major and micronutrient ionic forms and normal concentration range in the nutrient solution

Element	Ionic Form	Concentration in Solution (mg/L, ppm)
Major Elements		
Nitrogen (N)	NO_3^- or NH_4^+	100–200
Phosphorus (P)	HPO_4^{2-} or $H_2PO_4^-$*	30–50
Potassium (K)	K^+	100–200
Calcium (Ca)	Ca^{2+}	100–200
Magnesium (Mg)	Mg^{2+}	30–70
Micronutrients		
Boron (B)	H_3BO_3 or BO_3^{3-}**	0.2–0.4
Chloride (Cl)	Cl^-	5.0
Copper (Cu)	Cu^{2+}	0.01–0.1
Iron (Fe)	Fe^{2+} or Fe^{3+}	2–12
Manganese (Mn)	Mn^{2+}	0.5–2.0
Molybdenum (Mo)	MoO_4^{2-}	0.05–0.2
Zinc (Zn)	Zn^{2+}	0.05–0.10

 * Form depends on pH of the nutrient solution.
** It is being increasingly suggested that boron exists in the nutrient solution as molecular H_3BO_3.

Source: Jones, 1983.

Table 17   Elemental composition of early nutrient solutions compared to Hydro-Flo

	mg/L (ppm)						
					Hoagland/	Hydro-Flo	
Sacks	Knops	Pfeffers	Crones	Arnon	ALC-	ALC-	
Element	1860	1865	1900	1902	1950	BLC	BLC-1
Nitrogen (N)	140	164	164	140	210	132	162
Phosphorus (P)	100	46	46	50	31	58	58
Potassium (K)	386	134	234	386	234	200	284
Calcium (Ca)	341	195	195	170	200	136	136
Magnesium (Mg)	50	20	20	25	48	47	47
Iron (Fe)	trace	trace	trace	25	5	4	4

Source: Wilcox, 1983.

Table 18   Element concentrations for several "standard" solutions used in soilless culture

	mg/L (ppm)					
Element	Hoagland and Arnon	Cooper	Modified Steiner	Wilcox 1	Wilcox 2	Wilcox 3
Major Elements						
Nitrogen (N)	210	200	171	132	162	175
Phosphorus (P)	31	60	48	58	58	65
Potassium (K)	234	300	304	200	284	400
Calcium (Ca)	200	170	180	136	136	197
Magnesium (Mg)	48	50	48	47	47	44
Micronutrients						
Boron (B)	0.5	1.5	0.3	1.5	1.5	0.5
Copper (Cu)	0.02	0.1	0.2	0.1	0.1	0.05
Iron (Fe)	5.0	12.0	3.0	4.0	4.0	2.0
Manganese (Mn)	0.5	2.0	1.0	0.5	0.5	0.5
Molybdenum (Mo)	0.01	0.2	0.1	0.1	0.1	0.02
Zinc (Zn)	0.05	0.1	0.4	0.3	0.3	0.05

Source: Gerber, 1985.

It is clear from these tables that there exists a considerable range in element concentration in various nutrient solution formulas which is probably due to how the nutrient solution is to be used based on a particular growing system and crop. Systems of hydroponic/soilless growing are discussed in greater detail in Chapter 7.

Crop Requirement Adjustments

The species being grown will require modification in the composition and replacement scheduling of the nutrient solution (Schon, 1992). Some crops are more sensitive to particular elements than others. Therefore, one formula will work well for one crop but poorly for others. This can be seen in the levels of the major elements in the nutrient solution based on the crop as suggested by Agrodynamics (Table 19).

van Zinderen Bakker (1986) has specified nutrient formulas for tomato, lettuce, and rose, as given in Table 20.

In the two examples given in Tables 19 and 20, the growing technique for the crop-based nutrient solution compositions is not specified; however, Ames and Johnson (1986) have described their nutrient solution concentrations for lettuce and cucumber grown in rockwool, and Ingratta et al. (1985) have done so in an unidentified substrate for tomato and cucumber, as shown in Tables 21 and 22, respectively.

Table 19   Recommended major element nutrient solution levels by crop

	Major Elements (mg/L, ppm)				
Crop	Nitrogen	Phosphorus	Potassium	Calcium	Magnesium
Cucumber	230	40	315	175	42
Eggplant	175	39	235	150	28
Herbs	210	80	275	180	67
Lettuce	200	50	300	200	65
Melon	186	39	235	180	25
Pepper	175	39	235	150	28
Tomato	200	50	360	185	45

Source: Schon, 1992.

Table 20   Nutrient solution formulas for tomato, lettuce, and rose

Reagent (fertilizer grade)	g/1000 L		
	Tomato	Lettuce	Rose
Major Elements			
Calcium nitrate (15.5-0-0)	680	407	543
Magnesium sulfate	250	185	185
Potassium nitrate (13-0-44)	350	404	429
Potassium chloride (0-0-60)	170	—	—
Monopotassium phosphate (0-53-34)	200	136	204
Ammonium nitrate (33.5-0-0)	—	60	20
Micronutrients			
Iron chelate (10% Fe)	15.0	19.6	19.6
Manganese sulfate (28% Mn)	1.78	0.960	3.9
Boron (Solubor) (20.5% B)	2.43	0.970	1.1
Zinc sulfate (36% Zn)	0.280	0.552	0.448
Copper sulfate (25% Cu)	0.120	0.120	0.120
Sodium molybdate (39% Mo)	0.128	0.128	0.128

Source: van Zinderen Bakker, 1986.

Requirements for Use

It has only been in recent years that the aspect of nutrient solution concentration management has been fully appreciated (Schon, 1992), although systems of nutrient solution control have yet to be fully put into common use. However, it must be noted that growers over the years have been successful using Hoagland and Arnon-based nutrient solution formulas, due in part to the generally good results obtained for a wide range of crops and replacement schedules. It bears repeating that plants do well on dilute Hoagland and Arnon nutrient solutions, and such dilutions can be used to control the rate of growth for those crops with low elemental requirements.

The ratio of the concentration of elements in the nutrient solution can have a marked effect on plant absorption, as was discussed earlier. Therefore, concentration ratios may be as important, as Steiner (1980) suggests, as the absolute concentration of any one element. The proper adjustment of elemental ratios is particularly important in the relationship among the major elements and the ratio of cations to anions in the nutrient solution, which

Table 21 **Nutrient solution concentration for lettuce and cucumber in Grodan Rockwool**

Element	mg/L (ppm)	
	Lettuce	Cucumber
Major Elements		
Nitrogen-nitrate (N-NO$_3$)	200	150
Phosphorus (P)	60	35
Potassium (K)	300	300
Calcium (Ca)	170	150
Magnesium (Mg)	5	30
Micronutrients		
Boron (B)	0.3	0.2
Copper (Cu)	0.1	0.2
Iron (Fe)	3.0	1.0
Molybdenum (Mo)	0.2	0.03
Zinc (Zn)	0.1	0.2
pH	—	5.0–6.0
EC, mmhos*	—	2.0

* Electrical conductivity.

Source: Ames and Johnson, 1986.

affects elemental uptake as the plant itself tends to maintain intrinsic ion balances specific to itself. The importance of these balances becomes of far greater significance when elements are absorbed differentially from the nutrient solution due to both plant requirements and alterations in the nutrient solution resulting from different uptake patterns. Therefore, plant roots *see* an entirely different nutrient solution with repeated use until the nutrient solution is renewed and the full complement of elements in their proper ratios is again available. Such a cyclic pattern of changing nutrient solution composition is surely less than ideal for optimum plant growth and development.

Current thinking about hydroponic growing systems calls for the use of a flowing nutrient solution technique in which the composition of the nutrient solution is constantly maintained either by one-way passage of the nutrient solution through the rooting channel (an open system) or in a closed system by continued replenishment of the depleted elements after

Table 22   Composition of nutrient solutions for tomato and seedless cucumber crops in substrate

| Elements | mg/L (ppm) | |
	Tomato	Cucumber
Major Elements		
Nitrogen (N)		
as nitrate (NO_3)	650	700
as ammonium (NH_4)	10	10
Phosphorus (P)	49	49
Potassium (K)	280	240
Calcium (Ca)	150	140
Magnesium (Mg)	25	20
Sulfur (S)	80	33
Micronutrients		
Boron (B)	0.2	0.1
Copper (Cu)	0.03	0.03
Iron (Fe)	0.6	0.6
Manganese (Mn)	0.6	0.6
Molybdenum (Mo)	0.05	0.05
Zinc (Zn)	0.3	0.3

Source: Ingratta et al., 1985.

each use. The objective is to keep the elemental ion concentration in the nutrient solution constant, an approach that closely parallels the elemental ion environment in soil in which the soil solution is resupplied with ions as they are removed by root absorption (Lindsay, 1979; Barber and Bouldin, 1984). Experiments have shown that plants growing in this type of hydroponic system more closely follow the pattern of plant growth and development obtained in soil (Asher and Edwards, 1978a).

A similar effect can be obtained by growing plants in a large volume of agitated nutrient solution where the volume is sufficiently large that its elemental ion composition is not affected by ion absorption or loss of water by plant uptake—the result of transpiration. Naturally, such a system of growing would not be practical, and thus the flowing technique is used to accomplish the constancy.

The level of management will determine to what degree control of the composition of the nutrient solution will be required. For the grower satisfied with average plant and fruit production goals, precise control is not warranted. For the grower reaching for the maximum genetic potential of the crop, precise control of the nutrient solution and its composition becomes essential. In the greenhouse, where precise control of the environment is the practice (Edwards, 1994), the use and composition of the nutrient solution must be brought into the management decision-making process. Currently, the exact control requirements are not fully known, but research and experience are beginning to set these parameters (Bauerle et al., 1988; Bauerle, 1990).

Nutrient Solution Control

All systems of nutrient solution management, whether open or closed, must lend themselves to precise control of the nutrient solution composition so that the concentration of elements can be varied in response to both known physiologic stages of the developing plant and the grower's sense of the condition of the crop (Schon, 1992). When beginning, it is advisable to have the constituted nutrient solution assayed by means of a laboratory analysis. Such an analysis ensures that all of the elements are in the nutrient solution at their desired concentration.

In a closed recirculating hydroponic system, it is necessary to add water to the nutrient solution in order to maintain its original volume, which is very important. In addition, some of the elements will have been removed along with the water; these elements can be included in the make-up water. The question is how much of which elements should be added. A common practice is to use an electrical conductivity (EC) measurement of the nutrient solution as a means of determining what level of replenishment is needed; this technique works fairly well. Unfortunately, such a measurement does not determine what differential change in elemental concentration may have taken place in order to add back what was removed, element by element. Such a determination requires a complete elemental analysis of the solution.

The elements that are most likely to show the greatest change in the nutrient solution with use are nitrogen and potassium. A possible rule of thumb would be to dilute the initial nutrient solution formula for the major elements only and add that as the make-up water, making this solution about one-quarter to one-third the strength of the original nutrient solution. Some experimenting and testing will be necessary to determine what that proper strength should be to avoid creating an ion imbalance by adding back too much or too little. The micronutrients should never be included in the make-up nutrient solution in order to minimize the possible danger from excesses. Phosphorus is also an element that should possibly be excluded from the make-up solution. Recommendations for the composition of make-up water will be given later for particular hydroponic systems.

Another factor that must be considered is what elements were being left behind in the rooting media; the amount will vary depending on the media characteristics, the composition of the nutrient solution, and the frequency of recirculation. An important measurement that is recommended with some growing systems is to periodically take an aliquot of solution from the rooting media and determine its EC. At some designated EC reading, the rooting media would then be leached with water to remove accumulated *salts*.

Anyone who has used gravel as a rooting medium, for example, may have noticed that with time a grey-white sludge (primarily precipitated calcium phosphate and sulfate) begins to appear which may also entrap other elements, particularly the micronutrients. The sludge can be a major source of elements for plant uptake irrespective of what is being added by means of the nutrient solution. This accumulation of sludge and its utilization by the plant can give rise to a gradual or sudden marked change in plant elemental content, which frequently is undesirable. Therefore, control of this type of accumulation needs to be part of the nutrient solution management program.

pH

The pH of the nutrient solution is thought to be best when kept between 6.0 and 6.5, although most nutrient solutions, when initially constituted, will have a pH between 5.0 and 6.0. It is well known that if the nutrient solution pH drops below 5.0 or goes over 7.0, plant growth may be affected. Ikeda and Osawa (1981) observed that 20 different vegetable species showed a similar nitrogen source preference for either nitrate (NO_3^-)- or ammonium (NH_4^+)-nitrogen when the pH of the nutrient solution was varied from 5.0 to 7.0. Similar experiments need to be conducted for the other essential elements to determine the effect of nutrient solution pH on elemental uptake by plants.

The pH of the nutrient solution markedly affects the availability of certain elements, particularly the micronutrients, stimulating excessive uptake at low pH and resulting in removal from the nutrient solution by precipitation at high pH. Therefore, careful control of the pH is important in order to both keep all the essential elements in solution and prevent toxicity due to excessive uptake. Another means of control to minimize the pH and other effects is to use the chelated forms of the micronutrients; forms and concentration levels have been suggested by Wallace (1971, 1989).

It is believed that the pH of the nutrient solution is less critical in a flowing solution culture system than in one that is static, as long as the pH remains between 5.0 and 7.0. Therefore, control of pH in flowing nutrient solution systems is substantially less demanding.

A considerable degree of pH control can be obtained by simply selecting a specific ratio of nitrate (NO_3^-) to ammonium (NH_4^+) ions when the nutrient solution is initially prepared. If the ratio of NO_3^- to NH_4^+ is greater than 9 to 1, the pH of the solution tends to increase with time, whereas at ratios of 8 to 1 or less, pH decreases with time, as illustrated in Figure 4.

Appropriate combinations of either mono- or dihydrogen phosphate salts (HPO_4^{2-} and $H_2PO_4^-$, respectively) of either calcium or potassium will also give some degree of pH control with time.

Diurnal fluctuations in pH occur as the result of the changing solubility of carbon dioxide in the nutrient solution; however, these changes are usually not of sufficient magnitude to warrant daily adjustment.

The pH of the nutrient solution can be adjusted by monitoring the pH continuously or adding an acid or alkali, as the case requires, to either lower or raise the pH after each period of use. The common procedure is to continuously monitor the pH and inject acid or alkali into the flowing stream of nutrient solution. Solutions of either sodium or potassium hydroxide (NaOH and KOH, respectively) are suitable alkalis for raising the pH. Ammonium hydroxide can also be used; however, it is more difficult to handle safely, and the addition of the ammonium ion to the nutrient solution may not be desirable. Nitric (HNO_3), sulfuric (H_2SO_4), and hydrochloric (HCl) acids can be used for lowering the pH. Phosphoric acid (H_3PO_4) can also be used, but its use would add phosphorus, which might not be desirable.

Those acids and alkalis that contain one or more of the essential elements are less desirable for use than those that do not contain such elements. Thus, NaOH is the preferred alkali, and either H_2SO_4 or HCl is the preferred acid. Commercially available pH control solutions for use in nutrient solutions are usually made from these reagents.

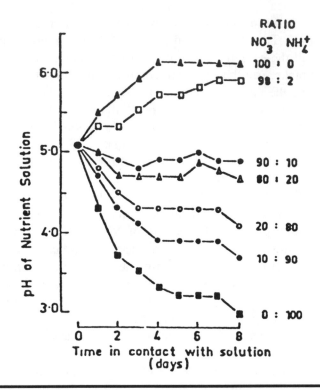

Figure 4   Effect of the ratio of nitrate to ammonium nitrogen on the rate and direction of pH changes in nutrient solutions in contact with the roots of wheat (*Triticum aestivium*) plants. (Source: Trelease and Trelease, 1935.)

As mentioned earlier (Chapter 4), some plants will effectively reduce the pH of the solution in the immediate vicinity of their roots. This acidification enhances their ability to absorb certain elements, such as iron (Rodriguez de Cianzio, 1991). If the nutrient solution is constantly being adjusted upward to a neutral pH, it can interfere with the plant's natural ability to enhance its elemental ion-absorptive capability. Therefore, some have suggested that the pH of the nutrient solution should not be continuously adjusted but instead should be allowed to seek its own level naturally. This may be a desirable practice with those plant species sensitive to iron when being grown hydroponically.

pH control of the nutrient solution may be akin to nutrient solution filtering, discussed earlier. It may be that more has been made about pH control and its potential effect on plants than can be justified from actual experience. Therefore, the requirement for pH control becomes a management decision, balancing benefits gained versus costs to control. It is obvious that there are extremes which the pH of the nutrient solution should not be allowed to reach. What is needed to maintain the pH and prevent it from reaching those extremes may be academic, since those extremes are seldom reached within normal experience with most nutrient solution formulas and their requirements.

Temperature

The temperature of the nutrient solution should never be less than the ambient air temperature, particularly in systems where plant roots are exposed to intermittent surges of a large volume of nutrient solution. On warm days, when the atmospheric demand on plants is high, root contact with nutrient solution below the ambient temperature results in plant wilting—an undesirable stress on plants. Plant roots sitting in cool or cold nutrient solution cannot absorb sufficient water and elements to meet the demand imposed by warm air and bright sunshine. Repeated exposure to cool nutrient solution results in plant growth and performance below expected levels, evidenced by poor fruit set and quality and delayed maturity. In such circumstances, it may be necessary to warm the nutrient solution to avoid this stress. On the other hand, warming the nutrient solution above the ambient temperature is not recommended and may do harm to the crop.

Electrical Conductivity

Most nutrient solution formulas have a fairly low [<3.0 mmhos/cm (µS/cm)] electrical conductivity (EC) when initially made. For example, the Hoagland No. 1 nutrient solution given in Table 11 has an EC of 2.7 mmhos/cm. The "salt effect" in a nutrient solution formulation can be minimized by selecting those compounds that have low salt indices (Table 23) when making the nutrient solution. It is with use and/or reuse that a soluble salt problem arises. This problem develops when substantial quantities of water are removed at a very rapid rate from the nutrient solution when in contact with plant roots, as happens on hot, low-humidity days. If the water loss is not replaced, the EC of the nutrient solution will rise; this becomes particularly acute if the nutrient solution is being recirculated and the water loss due to evapotranspiration is not immediately replaced.

Table 23   Relative salt index for common chemical reagents used for preparing nutrient solutions

Reagent	Formula	Relative Salt Index
Ammonium nitrate	NH_4NO_3	104
Ammonium sulfate	$(NH_4)_2SO_4$	69
Calcium nitrate	$Ca(NO_3)_2 \cdot 4H_2O$	52
Calcium sulfate	$CaSO_4 \cdot 2H_2O$	8
Diammonium phosphate	$(NH_4)_2HPO_4$	29
Monoammonium phosphate	$NH_4H_2PO_4$	34
Monocalcium phosphate	$CaHPO_4$	15
Potassium chloride	KCl	116
Potassium nitrate	KNO_3	73
Potassium sulfate	K_2SO_4	46
Sodium nitrate	$NaNO_3$	100

An EC measurement of the nutrient solution can also be used to determine the nutrient element replenishment level required to reconstitute the solution before reuse. From previous determinations, the amount of replenishment solution required to be added to the nutrient solution would be based on that EC measurement. Although this system of nutrient solution management has worked reasonably well, it does not take into account individual losses of elements from the nutrient solution by root absorption or retained in rooting media. Therefore, replenishment based on an EC measurement may not fully reconstitute the nutrient solution in terms of its elemental composition.

In rockwool and perlite bag culture, for example, measurement of the EC of an aliquot of the retained solution in the rockwool slab or perlite or the effluent from them can be used to determine when leaching would be required to remove accumulated salts.

Methods and Timing of Nutrient Solution Delivery

The timing and techniques for delivering the nutrient solution to the roots of plants will play a role in determining its composition. For a nutrient solution application schedule based primarily on the demand of the crop for water, the grower may be applying a nutrient solution when the nutrient demand by the plants is already satisfied—no additional nutrient elements at that specific time are needed. However, it is not common practice to just apply pure water to the plants, although such a capability would be desirable.

As discussed earlier, with increasing frequency of application of a nutrient solution, the concentration of the nutrient elements in solution should be lower (see Table 15). One could argue that on high atmospheric demand days when plants are rapidly transpiring, both water and nutrient element requirements would be near equal in terms of what is being supplied by the nutrient solution (assuming the nutrient solution formula is specifically made for these conditions, the crop, etc.), while on less demanding days, the requirement for both is less but still equal. By not adjusting the nutrient solution composition, it is assumed that the proper balance is achieved between water and nutrient element demand. Experience has shown that this concept is reasonably correct under most conditions. However, as the factors that relate to plant growth and development are better controlled, this assumed equal relationship between water demand and nutrient element need probably does not hold.

The size of the root mass is also a major factor that will affect water and nutrient element absorption (Barber and Bouldin, 1984). As the root surface area increases, the influx of water and nutrient elements through the roots also increases. In hydroponic systems, one might ask how large the root mass must be to ensure that the demand for water and nutrient elements is met. Unfortunately, no one has adequately made such a determination. There is some evidence which suggests that the root mass is not as important as root activity and that a large root mass may actually be detrimental to best plant growth and development.

The most common method of nutrient solution delivery used today with bag and slab culture hydroponic systems is by means of drip irrigation, which provides an intermittent delivery of the nutrient solution at the base of the plant, as shown in Figure 5.

Based on a predetermined schedule, nutrient solution flows from its reservoir out the end of the dripper; the frequency and rate of flow are adjusted to stage of plant growth, atmospheric conditions, etc. When the dripper is on, the area around the point of delivery is saturated with nutrient solution; when off, the nutrient solution drains away, creating a changing root environment that may not be best for optimum plant growth and development. With the draining of nutrient solution away from the point of introduction, air is drawn into the rooting medium, bringing in oxygen (O_2). Usually sufficient nutrient solution is applied so that the area immediately around the dripper is leached, pushing any unused accumulated nutrient

Figure 5   Placement of the drip tube shown at the base of the rockwool block.

elements deeper into the bag or slab. Normally, the bottom of the bag or slab is open, which allows excess nutrient solution to flow out. Based on an analysis (usually a determination of the EC of a drawn solution sample from the medium), water will be periodically applied through the dripper to leach the growing medium of any accumulated salts—the remaining unused nutrient elements.

In ebb-and-flow nutrient solution systems, the nutrient solution is pumped from a reservoir into the growing medium, flooding it with solution for a short period, and then the nutrient solution is allowed to flow out of the rooting medium back into the reservoir, as illustrated in Figure 6.

This outflow of nutrient solution from the growing medium draws air into the rooting bed, providing a source of oxygen. From the moist rooting medium, plants are able to obtain water and nutrient elements. Again, in such a system of nutrient solution delivery, the roots experience a changing environment, which may not be ideal for best plant growth and development, although plant performance is usually satisfactory. In the rooting medium (sand and gravel are two common materials used for such systems), an accumulation of unspent nutrient elements will occur as a precipitate as the medium dries out; the precipitate is primarily a mixture of calcium phosphate and sulfate, which will also occlude other elements. Periodically, the medium will require intensive leaching or replacement, as the accumulating precipitate will begin to significantly affect plant growth and development.

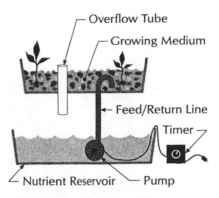

Figure 6   A typical ebb-and-flow hydroponic system. (Source: *The Best of the Growing Edge,* 1994, p. 9.)

For standing aerated systems, roots are suspended in a continuously aerated nutrient solution. Depending on the volume of nutrient solution versus number of plants, the nutrient solution will be significantly changed and will require periodic replenishing.

In the nutrient film technique, the nutrient solution flows down a channel occupied by plant roots. As the distance from the point of introduction increases, the characteristics of the nutrient solution will significantly change; first the dissolved oxygen in the nutrient solution dissipates (Antkowiak, 1993), and a change in the elemental composition of the solution also occurs. Therefore, the length of flow is critical. As the root mass increases, the nutrient solution will tend to flow over the root mass rather than through it, which will significantly affect plant performance with time.

For the aeroponic system, the nutrient solution periodically bathes the roots with a fine mist of nutrient solution; the finer the mist, the better the plant performance. Oxygen deficiency is not a problem, but the frequency of misting must be sufficient to keep the roots supplied with enough water to meet the transpiration demand of the plant.

None of these commonly used nutrient solution delivery systems is without some undesirable aspect, although all are capable of delivering sufficient water and essential elements to the plant. The question is which system will work best in terms of the efficient use of water and nutrient elements, resulting in high plant performance. The answer at this time is that none of them will, and the ideal delivery and utilization system has yet to be devised.

Systems of Hydroponic/ Soilless Culture

8

There are a number of ways to grow plants by means of hydroponic/soilless culture. For purposes of this book, the classification scheme offered by Dr. John Larsen of the Texas Agricultural Extension Service is followed. In his classification system, hydroponics is one distinct technique for plant growing where no root-supporting medium is used, whereas the other systems employ a rooting medium, either inorganic or organic. Larsen's classification system is given in Table 24; it has been modified to include several rooting media materials in common use today.

Table 24   Dr. John Larsen's hydroponic/soilless culture classification system

| Hydroponics or Water Culture | Media Culture | | |
	Inorganic	Organic	Mixtures
1. Standing aerated	1. Gravel	1. Peat moss	1. Peat moss/vermiculite/ perlite
2. Nutrient film technique	2. Sand	2. Pine bark	2. Pine bark/vermiculite/ perlite
3. Aeroponics	3. Perlite	3. Sawdust	3. Peat moss/pine bark/ perlite
	4. Rockwool		

As indicated earlier, growing plants hydroponically is different for systems that employ a support or rooting media compared to non-media systems. Management of the nutrient solution for these two classes of systems is quite different. It is important, however, to keep in mind not only the differences but also the similarities between these growing systems, as some of the management procedures can be successfully transferred, whereas others cannot.

Container Growing

All forms of hydroponic/soilless culture involve growing plants in some kind of a container—a bed, pot, bag, can, enclosed slab, or trough. The volume and dimensions of the rooting vessel are frequently chosen on the basis of convenience or availability. For example, in the past, the Number 10 food was widely used because of its low cost and availability. More recently, the so-called "gallon" and "2-gallon" (actual volumes are 3 and 6 quarts, respectively) containers have become popular. Today, growers are placing soilless medium in a free-standing plastic bag and using it as the growing container, or they are growing directly in the bag that is used to package and transport a soilless mix or perlite.

What should the volume and dimensions for the rooting vessel be, whether the vessel is a bag, slab, pot, trough or bed, in order to provide adequate space for normal root growth and development? The answer to that question, as far as most hydroponic/soilless growing systems are concerned, has not been adequately determined. It is surprising how little good information is available on the importance of rooting volume required by plants and the relationship that exists between rooting habit, rooting media, and container environment and volume. A brief discussion of roots and their effect on plant growth and development was presented in Chapter 4.

Despite the uncertainties about the relationship between rooting vessel size and plant performance, there are some guidelines that will assist the grower in determining the rooting volume needed for the crop and system being employed:

1. For all containers, the depth should be one-and-a-half to two times the diameter of the surface area covered by the plant canopy when the plant reaches its maximum size. For example, if the canopy covers (or will cover) a surface area 12 inches in diameter, the growing container should be 18 to 24 inches deep.

2. In bed culture systems, increased spacing between plants can, in part, substitute for a lesser depth. For example, plants with a canopy occupying a surface area 12 inches in diameter growing in a bed less than 12 inches deep should be spaced 18 inches from one plant center to another. This ratio of 2 to 3 can be applied to plants with smaller or larger canopies when growing in bed systems.

It is generally accepted that roots of neighboring plants inhibit each other's growth. Therefore, close contact and intermingling of roots between neighboring plants (the result of close spacing or shallow rooting depth) should be minimized by providing the proper area and depth required.

Some feel that the present lack of knowledge about root growth in varying environments restricts our knowledge of plant growth in general. I would agree! The consequence for the hydroponic/soilless culture grower is that he or she must experiment with the growing system to determine the rooting volume required to obtain maximum plant performance. Beginning with the recommendations given above, plants can be spaced closer together until a significant change in plant growth and yield appears.

Needless to say, root volume requirement becomes academic when plants must be widely spaced to allow sufficient light to penetrate the plant canopy for those plants that are widely branched and/or grow tall.

Media Hydroponic/Soilless Culture

From 1930 to the late 1950s, gravel or sand was commonly used as the rooting medium in closed recirculating ebb-and-flow commercial soilless culture systems. For small home hydroponic units, gravel, lava rock, or Hadite is the material selected for use as the rooting medium. For the commercial hydroponic systems of today, perlite and rockwool are the most commonly used inorganic rooting media materials.

A wide variety of various organic rooting media materials is used today, most of which are combinations of various materials, primarily mixtures containing peat moss and/or pine bark or peat moss and pine bark mixed with inorganic substances, such as vermiculite and perlite. The formulations in common use today are presented in detail in Chapter 10.

The use of a rooting medium, whether inorganic or organic, poses a set of challenges. Although the medium itself may be inert, such as gravel, sand, perlite, rockwool, etc., it harbors pore spaces that will hold nutrient solution which may eventually be absorbed by plant roots; the elements move with the solution by mass flow or by diffusion within the solution and are also reached by root extension (growth) (see Chapter 3). Organic media, such as peat moss and pine bark, have similar pore spaces, as well as a cation/anion exchange capacity that can remove ions from the solution and hold them for later release into solution. In both types of media, a precipitate of elements can occur, essentially as a combination of calcium phosphate and sulfate, which can also entrap other elements, mainly the micronutrients. Although this precipitate is essentially insoluble, portions can become soluble, which will then contribute to the essential element supply being delivered to the plant roots by repeated passage of the nutrient solution through the rooting medium.

Regulating Water and Nutrient Element Requirements

There are two basic systems of nutrient solution use:

- An open system in which the nutrient solution is passed through the rooting vessel and discarded
- A closed system in which the nutrient solution is passed through the rooting vessel and then collected for reuse

There are advantages and disadvantages to both systems. The major disadvantage to the open system is its inefficiency due to the loss of water and unused essential elements, since the flow of the nutrient solution is greater than that required by the plants. For the closed system, the nutrient solution can be substantially changed when passed through the rooting vessel, requiring some adjustment in volume (replacement of lost water) and pH and replenishment of absorbed essential elements (Hurd et al., 1980). In addition, any disease or other organisms picked up by the nutrient solution in its passage through the rooting vessel will be recirculated into the entire system unless removed or inactivated by some form of nutrient solution treatment. The controls and requirements for a recirculating hydroponic system have been discussed by Wilcox (1991), Schon (1992), and Bugbee (1995).

The nutrient solution is expected to provide both water and the essential elements needed by the plant in its flow through the rooting vessel. It is easily and erroneously assumed that these two physiological requirements, the need for both water and essential elements, occur in tandem. On warm days when plants are transpiring rapidly, only water may be needed to meet the atmospheric demand, while the nutrient elements in the nutrient solution may not be required by the crop in other than their usual amounts. The consequence is that the need for water is out of phase with the feeding cycle. This juxtaposition of events poses a major problem, as it is not common to have a water-only system operating in parallel with the nutrient solution delivery system. Therefore, increasing the circulation of the nutrient solution to meet the demand for water may lead to an elemental imbalance and an undesirable accumulation of unwanted elements.

With automatic control (Bauerle et al., 1988; Berry, 1989; Bauerle, 1990; Edwards, 1994) and an open system, it is possible to modify the nutrient solution composition by adding water into the flowing stream of nutrient solution passing through the rooting vessel, thereby reducing the nutrient element concentration. With a closed system, a delivery–collection system would be required to pass water only through the rooting vessel. Such "engineering" aspects of hydroponic culture have recently been discussed by Giacomelli (1991).

Active and Passive Systems of Nutrient Solution Distribution

In all commercial and most other types of hydroponic/soilless culture systems, the movement of the nutrient solution requires either electrical power (active) or gravity (passive), or a combination of both. For some situations, less dependency on electrical power can be of considerable advantage. However, with the requirements for greater control over the composition, application requirements, etc. of the nutrient solution being more widely recommended and applied in commercial systems (Bauerle et al., 1988; Berry, 1989; Bauerle, 1990; Schon, 1992), the need for uninterrupted electrical power is becoming essential. In addition, computer programmed systems are replacing manual management operations. Sensors are being placed in the growing medium and nutrient solution storage tanks for regulating the flow and composition of the nutrient solution, respectively. Measurements such as light intensity and duration and the temperature of the plant environment are factors being used to regulate the flow and composition of the nutrient solution. Therefore, it would seem that passive systems of nutrient solution flow are becoming obsolete.

Systems of Hydroponic Culture

9

According to the classification system given by Larsen (see Table 24), true hydroponics is the growing of plants in nutrient solution without a rooting medium. Plant roots are either suspended in standing aerated nutrient solution or in a nutrient solution flow through a root channel or plant roots are sprayed periodically with a nutrient solution. This definition is quite different from the usually accepted concept of hydroponics, which has in the past included all forms of hydroponic/soilless growing. In this chapter, these three techniques of hydroponic growing will be discussed, as well as hydroponic systems using inorganic rooting media.

True Hydroponic Systems (Mediumless)

Standing Aerated Nutrient Solution

This is the oldest hydroponic technique, dating back to those early researchers who, in the mid-1800s, used it to determine which elements were essential for plants. Sachs in the 1840s and the other early investigators grew plants in aerated solutions and observed the effect on plant growth with the addition of various substances to the solution (Russell, 1950). This tech-

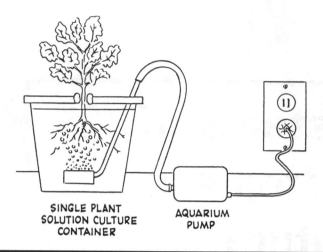

SINGLE PLANT
SOLUTION CULTURE
CONTAINER

AQUARIUM
PUMP

Figure 7   Standing aerated nutrient solution hydroponic system.

nique is still of use for various types of plant nutrition studies, although some researchers have turned to flowing and continuous replenishment nutrient solution procedures.

The requirements for the aerated standing nutrient solution technique are (1) a suitable vessel, (2) a nutrient solution, and (3) an air tube which bubbles air continuously into the nutrient solution, as shown in Figure 7. The bubbling air serves to add oxygen (O_2) to the nutrient solution as well as stir it. Depending on the size and number of plants and volume of nutrient solution, the nutrient solution is replaced on a predetermined schedule, usually between 7 to 14 days. Water loss from the solution must be replaced daily, usually with pure water. The commonly used formula is Hoagland's (see Table 11) or some modification of it as has been designed by Berry (1985), whose nutrient solution formula is given in Table 25. The plant/nutrient solution volume ratio is 1 plant per 2 to 4 gallons (9 to 18 L) of nutrient solution.

Another aerated standing nutrient solution system has been described by Clark (1982); the technique is used for studies on the elemental requirements of corn and sorghum. Several plants are grown in $1/2$ gallon (2 L) of nutrient solution, with change schedules varying from 7 to 30 days depending on the stage of growth and plant species. The ratio of nitrate (NO_3^-) to ammonium (NH_4^+) in the nutrient solution is used to control the pH; the ratio

Table 25 Preparation of stock concentrates for nutrient solution (200:1 dilution)

Reagent	Formula	g/L	oz./5 gallons
Stock Concentrate #1			
Potassium nitrate	KNO_3	50.55	33.8
Potassium phosphate (mono)	KH_2PO_4	27.22	18.2
Magnesium sulfate	$MgSO_4 \cdot 7H_2O$	49.30	32.9
Micronutrient concentrate		100 mL	64 fl. oz.
Micronutrient Concentrate			
Boric acid	H_3BO_3	2.850	1.90
Manganese sulfate	$MnSO_4 \cdot H_2O$	1.538	1.03
Zinc sulfate	$ZnSO_4 \cdot 7H_2O$	0.219	0.15
Copper sulfate	$CuSO_4 \cdot 5H_2O$	0.078	0.05
Molybdic acid	$MoO_2 \cdot 2H_2O$	0.020	0.01
Stock Concentrate #2			
Calcium nitrate*	$Ca(NO_3)_2 \cdot 4H_2O$	118.0	78.8
Sequestrene 330 Fe**		5.0	3.3

* Norsk Hydro Calcium Nitrate is used with the formula [$5Ca(NO_3) : 2NH_4NO_3 : 10H_2O$]; add only 88.8 g/L or 59 oz./5 gallons.
** Mix the iron chelate thoroughly in a small amount of water before adding to the calcium nitrate.

Approximate concentration of elements in final solution (mg/L, ppm):

Major Elements: NO_3-N = 103, PO_4-P = 30, K = 140, Ca = 83, Mg = 24, SO_4-S = 32

Micronutrients: B = 0.25, Cu = 0.01, Fe = 2.5, Mn = 0.25, Mo = 0.005, Zn = 0.025

Source: Berry, 1985.

Table 26 Composition of nutrient solution for standing aerated growing system

Solution Number	Stock Solution* Reagent	Concentration (g/L)	Solution Used (mL/L)	Full-Strength Nutrient Solution (mg element/L) Cation	Anion
1	$Ca(NO_3)_2 \cdot 4H_2O$	270.0	6.6	Ca = 302.4	NO_3-N = 211.4
	NH_4NO_3	33.8		NH_4-N = 39.0	NO_3-N = 39.0
2	KCl	18.6	7.2	K = 70.2	Cl = 63.7
	K_2SO_4	44.6		K = 142.2	S = 58.3
	KNO_3	24.6		K = 68.5	NO_3-N = 24.5

Table 26   Composition of nutrient solution for standing aerated growing system (continued)

Solution Number	Stock Solution* Reagent	Concentration (g/L)	Solution Used (mL/L)	Full-Strength Nutrient Solution (mg element/L) Cation	Anion
3	$Mg(NO_3)_2 \cdot 6H_2O$	142.4	2.8	Mg = 37.8	NO_3-N = 43.6
4	KH_2PO_4	17.6	0.5	K = 2.5	P = 2.00
5**	$Fe(NO_3)_3 \cdot 9H_2O$	13.31	1.5	Fe = 2.76	NO_3-N = 2.1
	HEDTA	8.68		Na = 4.48	HEDTA = 13.0
6	$MnCl_2 \cdot H_2O$	2.34	1.5	Mn = 0.974	Cl = 1.3
	H_3BO_3	2.04		B = 0.536	
	$ZnSO_4 \cdot 7H_2O$	0.88		Zn = 0.30	S = 0.147
	$CuSO_4 \cdot 5H_2O$	0.20		Cu = 0.076	S = 0.038
	$Na_2MoO_4 \cdot 2H_2O$	0.26		Na = 0.074	Mo = 0.155

Element	Final Composition mg/L (ppm)	µM
Calcium (Ca)	302	7,540
Potassium (K)	283	7,240
Magnesium (Mg)	37.8	1,550
Nitrate-nitrogen (NO_3-N)	321	22,900
Ammonium-nitrogen (NH_4-N)	39.0	2,780
Chlorine (Cl)	65.0	1,940
Phosphorus (P)	2.00	65
Iron (Fe)	2.76	49
Manganese (Mn)	0.974	18
Boron (B)	0.536	50
Zinc (Zn)	0.300	4.6
Copper (Cu)	0.076	1.2
Molybdenum (Mo)	0.155	1.6
Sodium (Na)	4.56	200
HEDTA	13.0	47

* In each solution, the respective reagents were dissolved together in the same volume. Some of the reagents in solutions 1 to 4 may be combined to make fewer stock solutions if desired, but calcium reagents should be kept separate from sulfate (SO_4) and phosphate (PO_4) reagents. Combinations of the salts noted are for convenience.

** This solution was prepared by (a) dissolving the HEDTA [N-2(hydroxyethyl)ethylene-diaminetriacetic acid] in 200 mL distilled water + 80 mL 1N NaOH; (b) adding solid $Fe(NO_3) \cdot 9H_2O$ to the HEDTA solution and completely dissolving the iron salt; (c) adjusting the pH to 4.0 with small additions of 1N NaOH in step (c) too rapidly to allow iron to precipitate. The HEDTA was obtained from Aldrich Chemical Co., Milwaukee, WI (Catalog No. H2650-2).

Source: Clark, 1982.

normally is 8 to 1 with a total of 300 mg/L (ppm) nitrogen in solution. Clark's nutrient solution formula is given in Table 26. Although Clark's technique is primarily designed for corn and sorghum nutritional studies, his method of nutrient solution management can be applied to other plant species.

This aerated standing nutrient solution method of hydroponic growing has limited commercial application, although lettuce and herbs have been successfully grown on styrofoam sheets floating on an aerated nutrient solution (Figure 8). The plants are set in small holes in the styrofoam, with their roots growing into the nutrient solution. The sheets are lifted from the solution when the plants are ready to harvest.

Another reason why this system of growing hydroponically is not well suited for commercial application is that water and chemical use are quite high due to frequent replacement. In addition, the constantly changing composition of the nutrient solution can impose a substantial burden on the grower in order to maintain proper and adequate elemental ion balance and sufficiency during the growth period, although lettuce is usually ready for harvest in less than 45 days. The more common procedure for lettuce production is by nutrient film technique (Gerber, 1986).

Nutrient Film Technique

A significant development in hydroponics occurred in the 1970s with the introduction of the nutrient film technique, frequently referred to as NFT (Cooper, 1976, 1979). Some have modified the name by using the word "flow" (Schippers, 1979) in place of "film," as the plant roots indeed grow in a flow of nutrient solution. When Allen Cooper first introduced his NFT system of hydroponic growing (1976), it was heralded as the hydroponic system of the future. It was, indeed, the first major change in hydroponic growing to be introduced since the 1930s. But experience has shown that NFT does not solve the common problems inherent in most hydroponic growing systems. However, this did not deter its rapid acceptance and use in many parts of the world, particularly in some parts of Western Europe and England. NFT has been widely talked about and tested (Khudheir and Newton, 1983; Hurd, 1985; Cooper, 1985, 1988; Edwards, 1985; Gerber, 1986; Molyneux, 1988; Hochmuth, 1991), but its future is highly questionable unless better means of disease and nutrient solution control are found. Cooper (1996) has just recently published a revision of his 1976 book on NFT.

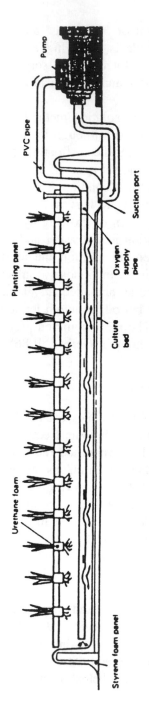

Figure 8   Lettuce plants set in styrofoam sheets floating on an aerated nutrient solution. (Source: Hydroponic Society of America Proceedings, 1989, p. 82.)

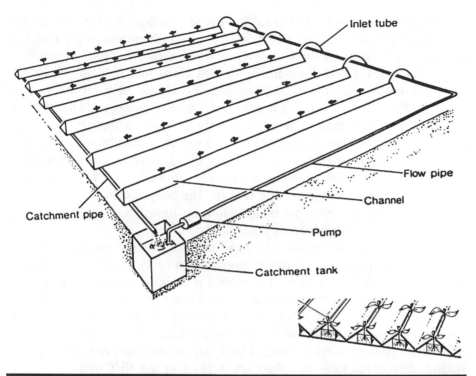

Figure 9   Allen Cooper's nutrient film technique (NFT) system. (Source: Cooper, 1979.)

Plant roots are suspended in a trough or channel through which the nutrient solution passes, as shown in Figure 9. Being a closed system, the nutrient solution is recycled in much the same manner as in gravel or sand bed-sump soilless media systems. Therefore, the difficulties associated with those systems can develop with NFT, except that there is no support medium which can absorb and retain elements from the nutrient solution.

The trough or channel containing the plant roots is set on a slope (usually about 1%) so that the nutrient solution can flow from the top to the lower end by gravity. Normally, the nutrient solution is introduced into the upper end of the trough at $1/4$ gallon (1 L) per minute. As the root mat increases in size, the flow rate down the trough diminishes. Plants at the upper end of the trough may reduce the oxygen (O_2) and/or elemental content of the nutrient solution sufficiently to significantly affect the growth and development of plants at the lower end. Furthermore, as the root mat thickens and becomes more dense, the flowing nutrient solution tends to move over the top and down the outer edge of the mat, reducing its contact with the entire root mass. This interruption in the flow results in poor mixing of the current flowing nutrient solution with water and elements left behind in the root mat from previous nutrient solution flows.

In order to minimize these effects, the trough should be no longer than 30 feet (9 m) in length and a minimum of 12 inches (30.5 cm) wide or wider, if possible. The channels are usually formed by folding a wide strip of polyethylene into a pipe-like trough. The polyethylene may be either white or black, but must be opaque to keep light out. If light enters the trough, algae growth becomes a serious problem.

The plants are set in the trough at the spacing recommended for that crop. Usually, plants are started in germination cubes made of fiberglass or similar material. The cube with its started plant is set directly in the trough. The polyethylene sheet is pulled around the plant stem and closed with pins or clips, forming a lightproof, pipe-like rooting trough (see Figure 9). Experience has shown that the germination cube should be of a substance other than peat or similar material that may disintegrate. A durable germination cube helps keep the plant set in place in the NFT trough.

The major advantages of NFT are the ease of establishment and the relative low cost of construction materials. Support for the trough can be inexpensively made of wood or sheet metal. The channel material, if made of strips of polyethylene, can be discarded after each crop, thus only necessitating sterilization of the permanent piping and nutrient solution storage tank.

Disease control can be difficult because a disease organism entering an NFT system will be quickly carried from one plant to another and one trough to another. Therefore, the same precautions are required as for any closed recirculating nutrient solution growing system. In warm climatic areas, the fungus *Pythium* is the major organism affecting plants grown in NFT systems. Presently, there seems to be no legal method of controlling this organism, although temperature and the concentration of copper, and possibly manganese and zinc, in the nutrient solution as well as its pH may be adjusted to offer some degree of control. However, current research and grower experience have not been sufficient to offer hope for effective *Pythium* control using any of these modifications of the nutrient solution. *Pythium* does not seem to be a serious problem when the temperature of the nutrient solution is maintained at less then 70°F (25°C).

Root death is another problem in NFT installations and may be the result of a lack of oxygen in the root mass (Antkowiak, 1993). Recently, it has been suggested that concern is greater than justified, inasmuch as root death is a natural physiological phenomenon brought on by competition within the plant for carbohydrates. During periods of high demand for carbohydrates (primarily at fruiting or during times of stress), some roots will die, but when stress is relieved, plant tissue regains an adequate carbohydrate supply and new roots will appear. As long as most of the roots in the mat are white in color, little attention should be paid to root death. This phenomenon probably occurs in all systems of growing but is clearly visible in NFT and not as easily seen in roots that are growing in an inorganic or organic medium.

A change in the design of the trough has been suggested by Cooper (1985), from the "U" shape to a "W" (called a divided gulley system) in which the plant sets on the top of the center and the roots are divided, as shown in Figure 10. A capillary matting is placed on the inverted "V" portion of the "W" to keep the roots moist with nutrient solution.

There are a number of advantages to this redesign of the NFT single-gully system as initially proposed by Cooper (1976, 1979). A portion of the

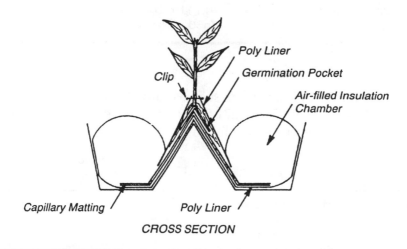

CROSS SECTION

Figure 10   Divided gully system ("W" design) devised by Cooper (1985).

rooting system—that on the inverted "V"—is in air; a portion of the roots lie on a moist surface (capillary matting), which provides for better oxygenation of the rooting system; and the remaining root mass is now divided into two channels, which should minimize the problems associated with a large mass of roots in a single channel. It is now possible to use two different irrigation systems by flowing water or various types of nutrient solutions down either channel. Unfortunately, the NFT channel system is now made more complicated in design, and it is uncertain whether this change will significantly improve plant performance.

Table 27   Nutrient solution formulas to give the theoretically ideal concentration of essential elements

Reagent	Formula	Amount (g/1000 L)
Potassium dihydrogen phosphate	KH_2PO_4	263
Potassium nitrate	KNO_3	583
Calcium nitrate	$Ca(NO_3)_2 \cdot 4H_2O$	1003
Magnesium sulfate	$MgSO_4 \cdot 7H_2O$	513
EDTA iron	$[(CH_2 \cdot N(CH_2 \cdot COOH)_2]_2FeNa$	79
Manganous sulfate	$MnSO_4 \cdot H_2O$	6.1
Boric acid	H_3BO_3	1.7
Copper sulfate	$CuSO_4 \cdot 5H_2O$	0.39
Ammonium molybdate	$(NH_4)_6Mo_7O_{24} \cdot 4H_2O$	0.37
Zinc sulfate	$ZnSO_4 \cdot 7H_2O$	0.33

Source: Cooper, 1979.

The theoretically ideal nutrient solution formula for the NFT system was given by Cooper (1979) (Table 27). This formula gives the following essential element concentrations in solution:

Element	Concentration (mg/L, ppm)
Major Elements	
Nitrogen (N)	200
Phosphorus (P)	60
Potassium (K)	300
Calcium (Ca)	170
Magnesium (Mg)	50

Element	Concentration (mg/L, ppm)
Micronutrients	
Boron (B)	0.3
Copper (Cu)	0.1
Iron (Fe)	12.0
Manganese (Mn)	2.0
Molybdenum (Mo)	0.2
Zinc (Zn)	0.1

Molyneux (1988) has also given NFT nutrient solution formulas, one for soft-water and another for hard-water use, which are shown in Table 28.

Table 28   NFT nutrient solution formulas for soft- and hard-water use

Reagent	Formula	Amount (kg/12.5 L)	
		Soft Water	Hard Water
Stock Solution A			
Calcium nitrate	$Ca(NO_3)_2 \cdot 4H_2O$	2.42	1.20
Ammonium nitrate	NH_4NO_3	—	60.0
Stock Solution B			
Potassium nitrate	KNO_3	1.53	2.59
Potassium dihydrogen phosphate	KH_2PO_4	0.55	—
Magnesium sulfate	$MgSO_4 \cdot 7H_2O$	1.27	1.27
		Amount (g/12.5 L)	
EDTA iron		75	75
Manganese sulfate	$MnSO_4$	10	10
Boric acid	H_3BO_3	6	6
Copper sulfate	$CuSO_4 \cdot 5H_2O$	2	2
Zinc sulfate	$ZnSO_4 \cdot 7H_2O$	1	1
Ammonium molybdate	$(NH_4)_6Mo_7O_{24} \cdot 4H_2O$	0.25	0.25

Note: Equal portions of Stock Solution A and B are mixed to make the nutrient solution. Phosphoric acid is used for pH adjustment to give a pH range of 6.0 to 6.5.

Source: Molyneux, 1988.

In addition, Molyneux (1988) has given the minimum, optimum, and maximum essential element concentrations in the NFT nutrient solution:

| | Concentration (mg/L, ppm) | | |
Element	Minimum	Optimum	Maximum
Major Elements			
Nitrate-nitrogen (NO_3-N)	50	150–200	300
Phosphorus (P)	20	50	200
Potassium (K)	50	300–500	800
Calcium (Ca)	125	150–300	400
Magnesium (Mg)	25	50	100
Micronutrients			
Boron (B)	0.1	0.3–0.5	1.5
Copper (Cu)	0.05	0.1	1.0
Iron (Fe)	3.0	6.0	12.0
Manganese (Mn)	0.05	1.0	2.5
Molybdenum (Mo)	0.01	0.05	0.1
Zinc (Zn)	0.05	0.1	2.5

Since NFT systems are operated as closed systems (i.e., the nutrient solution is recirculated a number of times before being discarded), Cooper (1979) has recommended the use of a special nutrient solution, referred to as the *topping-up solution*, to be added to the starting solution to maintain its composition during use. The starting solution and the topping-up formulas for tomato and cucumber are given in Table 29.

Table 29   Composition of nutrient solutions for the growing of tomato and cucumber in a NFT system

| | | Composition of Starting Solution | | |
Reagent	Formula	Stock Solution (g/L)	Dilution (mL/L)	Concentration (ppm)
Calcium nitrate	$Ca(NO_3)_2 \cdot 4H_2O$	787	1.25	117 N, 168 Ca
Potassium nitrate	KNO_3	169	3.9	254 K, 91 N
Magnesium sulfate	$MgSO_4 \cdot 7H_2O$	329	1.5	49 Mg
Potassium phosphate	KH_2PO_4	91	3.0	62 P, 78 K
Chelated iron	FeNaEDTA	12.3	3.0	5.6 Fe
Manganese sulfate	$MnSO_4 \cdot 4H_2O$	3.0	3.0	2.2 Mn
Boric acid	H_3BO_3	1.23	1.5	0.32 B
Copper sulfate	$CuSO_4 \cdot 5H_2O$	0.17	1.5	0.065 Cu
Ammonium molybdate	$(NH_4)_6Mo_7O_{24} \cdot 4H_2O$	0.06	1.5	0.007 Mo
Phosphoric acid	H_3PO_4	—	0.044	23 P

Table 29 Composition of nutrient solutions for the growing of tomato and cucumber in a NFT system (continued)

		Composition of Topping-up Solution		
Reagent	Formula	Stock Solution (g/L)	Dilution (mL/L)	Concentration (mg/L, ppm)
Calcium nitrate	$Ca(NO_3)_2 \cdot 4H_2O$	787	0.5*	47 N, 67 Ca
			1.0**	93 N, 113 Ca
Potassium nitrate	KNO_3	169	2.13	147 K, 51 N
Magnesium sulfate	$MgSO_4 \cdot 7H_2O$	329	1.0	32 Mg
Chelated iron	FeNaEDTA	24.5	0.4*	1.5 Fe
			0.8**	3.0 Fe
Manganese sulfate	$MnSO_4 \cdot 4H_2O$	7.42	0.3*	0.55 Mn
			0.6**	1.1 Mn
Boric acid	H_3BO_3	6.17	0.3	0.32 B
Copper sulfate	$CuSO_4 \cdot 5H_2O$	1.7	0.15	0.065 Cu
Ammonium molybdate	$(NH_4)_6Mo_7O_{24} \cdot 4H_2O$	0.06	1.5	0.007 Mo

 * For tomatoes.
** For cucumbers.

Source: Cooper, 1979.

Schippers' (1979) NFT nutrient solution formulas for the starting and topping-up nutrient solutions are given in Table 30.

Table 30 Nutrient solution formulations for the NFT system

		Starting Solution		Topping-up Solution	
Reagent	Formula	Weight (g/ 1000 L)	Conc. (mg/L)*	Weight (g/ 1000 L)	Conc. (mg/L)*
Calcium nitrate	$Ca(NO_3)_2 \cdot 4H_2O$	988.0	117 N	395.5	47 N
			168 Ca		67 Ca
Potassium nitrate	KNO_3	658.1	254 K	367.5	142 K
			91 N		51 N
Magnesium sulfate	$MgSO_4 \cdot 7H_2O$	496.6	49 Mg	324.3	32 Mg
Potassium phosphate	KH_2PO_4	272.0	62 P	—	—
			78 K	—	—
Chelated iron	FeNaEDTA	78.88	12 Fe	32.87	5 Fe

Table 30   Nutrient solution formulations for the NFT system (continued)

		Starting Solution		Topping-up Solution	
Reagent	Formula	Weight (g/ 1000 L)	Conc. (mg/L)*	Weight (g/ 1000 L)	Conc. (mg/L)*
Manganese sulfate	$MnSO_4 \cdot H_2O$	6.154	2 Mn	1.539	0.5 Mn
Boric acid	H_3BO_3	1.714	0.3 B	1.714	0.3 B
Copper sulfate	$CuSO_4 \cdot 5H_2O$	0.275	0.07 Cu	0.275	0.07 Cu
Ammonium molybdate	$(NH_4)_6Mo_7O_{24} \cdot 4H_2O$	0.092	0.05 Mo	0.092	0.05 Mo
Zinc sulfate	$ZnSO_4 \cdot 7H_2O$	0.308	0.07 Zn	0.308	0.07 Zn

* mg/L equivalent to parts per million (ppm).

Source: Schippers, 1979.

Other nutrient formulas have been proposed based on water source, whether soft (relatively free of ions such as calcium and magnesium) or hard (containing calcium and magnesium), and for various crops, such as tomato and cucumber. Such a set of formulas has been given by Papadopoulos (1991, 1994) as shown in Tables 31 to 34.

Table 31   Fertilizer formulations for use with NFT in soft-water areas

Stock Solution 1 (1000 L total volume)		Stock Solution 2* (1000 L total volume)		Stock Solution 3 (1000 L total volume)	
Reagent	Amount	Reagent	Amount	Reagent	Amount
Calcium nitrate	7.5 kg	Potassium nitrate	90 kg	Nitric acid (85%)	7.9 L
		Monopotassium phosphate	30 kg		
		Magnesium sulfate	60 kg		
		Iron chelate (15% Fe)	3.0 kg		
		Manganese sulfate	0.4 kg		
		Boric acid	0.24 kg		
		Copper sulfate	80 g		
		Zinc sulfate	40 g		
		Ammonium molybdate	10 g		

* It may be necessary to slightly acidify Stock Solution 2 with a small amount of nitric acid (20 mL) to prevent salt precipitation, e.g., magnesium phosphate.

Source: Papadopoulos, 1991.

Assuming a dilution ratio of 1:100 for Stock Solutions 1 and 2, the theoretical elemental concentrations in the circulating, diluted NFT solution are as follows:

Element	Concentration (mg/L, ppm)	Element	Concentration (mg/L, ppm)
Major elements		**Micronutrients**	
Nitrogen (N)*	214	Iron (Fe)	4.5
Phosphorus (P)	68	Manganese (Mn)	0.4
Potassium (K)	434	Boron (B)	0.2
Calcium (Ca)**	128	Copper (Cu)	0.09
Magnesium (Mg)	59	Zinc (Zn)	0.09
		Molybdenum (Mo)	0.09

* Additional nitrogen is supplied by the nitric acid of Stock Solution 3; however, the amount is small because the amount of acid needed to control the pH of soft water is far less than that required for hard water.

** The calcium content of the water supply has not been taken into account.

Table 32 Fertilizer formulations for use with NFT in hard-water areas*

Stock Solution 1 (1000 L total volume)		Stock Solution 2 (1000 L total volume)		Stock Solution 3 (1000 L total volume)	
Reagent	*Amount*	*Reagent*	*Amount*	*Reagent*	*Amount*
Calcium nitrate	50 kg	Potassium nitrate	80 kg	Nitric acid (67%)	54 mL
		Potassium sulfate	40 kg	Phosphoric acid (85%)	24 mL
		Magnesium sulfate	60 kg		
		Ammonium nitrate	0.6 kg		
		Iron chelate (15% iron)	3.0 kg		
		Manganese sulfate	0.4 kg		
		Boric acid	0.2 kg		
		Copper sulfate	80 g		
		Zinc sulfate	40 g		
		Ammonium molybdate	10 g		

* No phosphatic fertilizer has been included other than the phosphoric acid in Stock Solution 3. Where the water is not particularly hard and the acid requirement is correspondingly low, include 1.5 kg of monopotassium phosphate in Stock Solution 2 while decreasing the amount of potassium sulfate from 4.0 to 3.0 kg.

Source: Papadopoulos, 1991.

Assuming a dilution ratio of 1:100 for Stock Solutions 1 and 2, the theoretical elemental concentrations in the circulating diluted NFT solution are as follows:

Elements	Concentration (mg/L, ppm)
Major elements	
Nitrogen (N)*	192
Phosphorus (P)**	—
Potassium (K)	490
Calcium (Ca)***	85
Magnesium (Mg)	59
Micronutrients	
Boron (B)	0.4
Copper (Cu)	0.2
Iron (Fe)	4.5
Manganese (Mn)	1
Molybdenum (Mo)	0.5
Zinc (Zn)	0.09

  * Additional nitrogen is supplied by the nitric acid of Stock Solution 3.
 ** Some phosphorus is supplied by the phosphoric acid in Stock Solution 3.
*** The calcium content of the water supply has not been taken into account.

Table 33   Recommended nutrient solutions for tomato in NFT (amount of fertilizer per 1000 L of stock solution)

Stock Solution 1 (1000 L total volume)		Stock Solution 2 (1000 L total volume)	
Reagent	Amount	Reagent	Amount
Calcium nitrate	99.0 kg	Magnesium sulfate	49.7 kg
Potassium nitrate	65.8 kg	Monopotassium phosphate	27.2 kg
		Iron chelate (13% iron)	3.0 kg
		Manganese sulfate	0.5 kg
		Boric acid	180 g
		Copper sulfate	30 g
		Zinc sulfate	35 g
		Ammonium molybdate	8 g

Source: Papadopoulos, 1991.

- Prepare the final solution by adding equal volumes of both stock solutions in water until a recommended final solution electrical conductivity (EC) of 2200 μS/cm is achieved; adjust the pH to 6.2 by adding phosphoric (low-light conditions) or nitric (high-light conditions) acid. Ideally, stock solutions are mixed and pH is adjusted automatically by EC and pH controllers.
- When starting a new crop, begin with an EC of 1500 μS/cm and gradually increase to 2200 μS/cm over a week.
- A background EC of 300 to 600 μS/cm from the water supply is assumed.

Table 34   Recommended elemental levels for cucumber in NFT solutions

Stock Solution A (1000 L total volume)		Stock Solution B (1000 L total volume)	
Reagent	*Amount*	*Reagent*	*Amount*
Calcium nitrate	44.4 kg	Monopotassium phosphate	22.0 kg
Potassium nitrate	62.7 kg	Magnesium sulfate	50.0 kg
Ammonium nitrate	5.0 kg	Iron chelate (13% iron)*	1.0 kg
		Manganese sulfate (25% Mn)*	250.0 g
		Boric acid (14% B)*	90.0 g
		Copper sulfate (25% Cu)*	30.0 g
		Zinc sulfate (23% Zn)*	35.0 g
		Ammonium molybdate (57% Mo)*	8.0 g

* Alternatively, include 2.0 kg of Plant Product Chelated Micronutrient mix, which provides the following micronutrient concentrations (mg/L, ppm): 1.4 Fe, 0.4 Mn, 0.08 Zn, 0.26 B, 0.02 Cu, and 0.012 Mo.

- Prepare the final solution by adding equal volumes of both stock solutions in water until a recommended final solution EC of 2200 μS/cm is achieved; adjust the pH to 6.2 by adding phosphoric (low-light conditions) or nitric (high-light conditions) acid. Ideally, stock solutions are mixed and pH is adjusted automatically by EC and pH controllers.
- When starting a new crop, begin with an EC of 1500 μS/cm and gradually increase to 2200 μS/cm over 1 week.
- A background EC of 300 to 600 μS/cm from the water supply is assumed.
- The final dilution formula should give a NFT solution with the following elemental composition:

Element	Concentration (mg/L, ppm)
Major elements	
Nitrate (NO_3)	156
Ammonium (NH_4)	12
Phosphorus (P)	50
Potassium (K)	302
Calcium (Ca)	84
Magnesium (Mg)	50
Micronutrients*	
Iron (Fe)	1.3
Manganese (Mn)	0.62
Boron (B)	0.12
Copper (Cu)	0.07
Zinc (Zn)	0.08
Molybdenum (Mo)	0.03

* Alternatively, include 2.0 kg of Plant Product Chelated Micronutrient mix, which provides the following micronutrient concentrations (mg/L, ppm): 1.4 Fe, 0.4 Mn, 0.08 Zn, 0.26 B, 0.02 Cu, and 0.012 Mo.

Source: Papadopoulos, 1994.

In addition to the nutrient solution formula as given in Table 33, Papadopoulos (1991) has also described what the target elemental concentrations should be for tomato in a NFT system; those target values are listed in Table 35. A similar list of concentrations for tomato in a NFT system has been given by Ames and Johnson (1986), as shown in Table 36.

The influence of stage of plant growth is also a factor in determining what the elemental concentration ranges should be, as has been suggested by Hochmuth (1991) for the NFT technique for tomato; those ranges are shown in Table 37. As the stage of growth advances, there is an increase in the nitrogen, potassium, and magnesium concentrations, while the other elements remain at constant concentration.

Normally, the nutrient solution is monitored by periodic EC measurements which determine the appropriate times to add make-up (or topping-up) nutrient solution to maintain the initial volume and when to dump and make a new batch of nutrient solution.

The timing for flowing the nutrient solution down the NFT trough varies. One practice is to intermittently flow the nutrient solution down the

Table 35   Target elemental levels in NFT nutrient solutions for tomato cropping

Element	Minimum* (pH 5.5, EC 1800 µS)	Optimum (pH 6.0, EC 2000–2500 µS)	Maximum (pH 6.5, EC 3500 µS)
Major Elements			
Nitrogen-nitrate (N-NO$_3$)	50	150–200	300
Nitrogen-ammonium (N-NH$_4$)	5	10–15	20
Phosphorus (P)	20	50	200
Potassium (K)	100	300–500	800
Calcium (Ca)	125	150–300	400
Magnesium (Mg)	25	50	100
Sulfur (S)	—	50–200	
Micronutrients			
Boron (B)	0.1	0.3–0.5	1.5
Copper (Cu)	0.05	0.1	1.0
Iron (Fe)	1.5	6.0	12.0
Manganese (Mn)	0.5	1.0	2.5
Molybdenum (Mo)	0.01	0.05	0.1
Zinc (Zn)	0.05	0.5	2.5
Others			
Sodium (Na)	**	**	250
Chloride (Cl)	**	**	400

* Concentrations listed as minimal should be regarded as the approximate lower limit of a preferred range; in general, these minimum values are above those at which symptoms of deficiency develop. EC = electrical conductivity.
** As little as possible.

Source: Papadopoulos, 1991.

trough on an "on–off" cycle or by a "half-on, half-off" circulation period; a more sophisticated system is based on timing recirculation on the accumulation of incoming radiation. For example, when 0.3 mJ/m^2 of light energy has accumulated, the nutrient solution is flowed down the trough for 30 minutes; the time and length are also affected by the crop and its stage of growth. Such systems are coming into wider use because they have proven to be successful in producing better and higher yielding tomato and cucumber crops.

Table 36 Nutrient solution elemental concentrations for tomato in a NFT growing system

Element and Form	Concentration (mg/L, ppm)
Major Elements	
Nitrogen (N)	
nitrate (NO_3)	150–200
ammonium (NH_4)	0–20
Potassium (K)	300–500
Phosphorus (P)	50
Calcium (Ca)	150–300
Magnesium (Mg)	50
Micronutrients	
Boron (B)	0.3–0.5
Copper (Cu)	0.1
Iron (Fe)	3.0
Manganese (Mn)	1.0
Molybdenum (Mo)	0.05
Zinc (Zn)	0.1

Source: Ames and Johnson, 1986.

The NFT principle has also been applied to smaller growing units for home garden use. For example, one such application for vegetable growing places sand-filled styrofoam cups in access holes in PVC pipes; the nutrient solution circulates through the pipe on a timed schedule. This system has the unique feature of easy removal of plants by lifting the styrofoam cup from its access hole. A typical arrangement for this home garden NFT system is shown in Figure 11.

Aeroponics

Another promising hydroponic technique for the future was thought to be aeroponics, which is the distribution of water and essential elements by means of an aerosol mist bathing the plant roots. One of the significant advantages of this technique compared to flowing the nutrient solution past the plant roots is aeration, as the roots are essentially growing in air. The technique was designed to achieve substantial economies in the use of both water and essential elements. The critical aspects of the technique are the character of the aerosol, frequency of root exposure, and composition of the

Table 37 Final delivered nutrient solution concentrations for hydroponic (NFT-PVC pipe and rockwool) tomato in Florida greenhouses

	Stage of Growth* (mg/L, ppm)				
Element	1	2	3	4	5
Major Elements					
Nitrogen (N)	70	80	100	120	150
Phosphorus (P)	50	50	50	50	50
Potassium (K)	120	120	150	150	200
Calcium (Ca)	150	150	150	150	150
Magnesium (Mg)	50	50	50	60	60
Micronutrients					
Boron (B)	0.7	0.7	0.7	0.7	0.7
Copper (Cu)	0.2	0.2	0.2	0.2	0.2
Iron (Fe)	2.8	2.8	2.8	2.8	2.8
Manganese (Mn)	0.8	0.8	0.8	0.8	0.8
Molybdenum (Mo)	0.05	0.05	0.05	0.05	0.05
Zinc (Zn)	0.3	0.3	0.3	0.3	0.3

* Stage 1 = transplant to first cluster. Stage 2 = first cluster to second cluster. Stage 3 = second cluster to third cluster. Stage 4 = third cluster to fifth cluster. Stage 5 = fifth cluster to termination.

Source: Hochmuth and Hochmuth, 1996.

nutrient solution. Adi Limited (1982) described an aeroponic system which it said had proven to be highly successful; the system is computer controlled and requires a special fogging device, troughs, and an array of sensing devices. Although yields of crops obtained with this growing system have been reported to be considerably above those obtained with conventional hydroponic systems, the initial cost for the Adi system plus operating costs are very high, bringing into question its commercial viability (Soffer, 1985), although its value in plant propagation is considerable (Soffer, 1988).

Several methods have employed a spray of the nutrient solution rather than a fine mist; droplet size and frequency of exposure of the roots to the nutrient solution are the critical factors. Continuous exposure of the roots to a fine mist gives better results than intermittent spraying or misting. In most aeroponic systems, a small reservoir of water is allowed to remain in the bottom of the rooting vessel so that a portion of the roots has access to a continuous supply of water. The composition of the nutrient solution

Figure 11   **Nutrient flow system with styrofoam cups placed in holes in a PVC pipe of flowing nutrient solution. (Source: *Growing Edge*, 1993, p. 40.)**

would be adjusted depending on the time and frequency of exposure of the roots to the nutrient solution.

Medium Hydroponic Culture Systems

In the culture systems described in this section, plants are grown in some type of inorganic rooting medium (Straver, 1996a, b) with the nutrient solution applied by flooding or drip irrigation.

Ebb-and-Flow Nutrient Solution Systems

This type of hydroponic growing system had been in wide use for many years, although it is not commonly used commercially today other than for hobby/home-type growing units. The components are a growing bed containing an inert rooting medium, such as gravel, sand, volcanic rock, etc.; another container (sump) of equal volume of the growing bed, which contains the nutrient solution; and the appropriate piping, valves, and pump required to periodically flood the growing bed with nutrient solution which flows through a drain system from the growing bed back into the sump. Such a commercially designed system is shown in Figure 12.

The timing schedule for flooding the growing bed will depend on the atmospheric demand and stage of growth for the crop, as well as the water-holding capacity of the growing medium. Normally, the composition of the nutrient solution is similar to the basic Hoagland solution (see Table 11) or some modification of it, depending on the crop and stage of growth (see Chapter 7).

Being a closed system, the nutrient solution is recirculated until no longer suitable, when it is dumped and replaced with freshly made nutrient solution. Prior to each use, the nutrient solution should be tested for its pH, EC, and possibly its elemental content and then adjusted accordingly. The nutrient solution may also require filtering and sterilization after each circulation through the rooting bed. All of these procedures are discussed in Chapter 7.

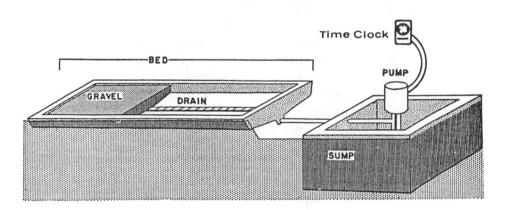

Figure 12   Commercial ebb-and-flow system for hydroponic growing.

This system of hydroponic growing has proven to be difficult to manage and is very inefficient in its use of water and essential elements, important reasons for its lack of use today. The rooting medium will require removal of the root mass and sterilization between crops, with periodic replacement required, which can be difficult and expensive. However, for the homeowner and hobbyist, this system of growing is relatively easy to construct and operate on a small scale and gives reasonably good plant performance with a moderate level of care.

Bag or Pot Drip/Pass-Through Nutrient Solution Systems

This system of hydroponic growing is in common use today for commercial production in which the plant(s) is grown in a bag or pot of medium; perlite is most commonly used (Gerhart and Gerhart, 1992). In one system, the bag used for shipping the perlite is laid on its side, small holes are cut along the bottom edge of the bag to allow excess water to flow out, an access hole(s) is cut in the top of the bag for placement of a plant, and then a drip tube is placed on the edge of the access hole next to the plant. Today, tomato and cucumber are two commonly grown crops using this bag culture system, as shown in Figure 13. A pot or other suitable container can be used in place of the shipping bag.

Being an open system, the nutrient solution is not recovered, and the amount delivered should be sufficient for a slight excess flow from the cut openings on the bottom edge of the bag or pot. Scheduling of the rate and timing of nutrient solution application is dependent on various factors, such as atmospheric demand, crop, stage of growth, etc. (see Chapter 7). During the growing period, the effluent from the bag can be monitored for its pH and EC and adjustments made in the nutrient solution delivered, or the bag can be leached with water to remove any accumulated salts. Also, an aliquot of solution can be drawn from the bag itself shortly after an irrigation to make the same measurements as made on an effluent sample.

At the end of the growing season, the perlite-containing bag may be used one more time or discarded, which makes the system relatively easy to install and replace at a reasonable cost. The nutrient solution formula normally is based on the Hoagland nutrient formula (see Table 11) or some modification of it (see Chapter 7).

Various modifications of this system of growing have been made to accommodate different types of crops. One example is a vertical hanging

Figure 13 Typical perlite bag culture system for hydroponic growing. (Courtesy of CropKing, Inc.)

bag with lettuce plants placed in holes in the side of the bag, a system described by DeKorne (1992–93). Another example is strawberry plants set in the holes in the side of the vertical polyethylene bag of perlite, as shown in Figure 14. The nutrient solution is applied at the top of the bag, usually through a dripper, and the solution passes down through the bag and out the bottom. The same problems associated with the NFT technique apply to this system, as the composition of the nutrient solution is modified as it passes down through the bag.

A very recent unique system consists of a column of interlocking styrofoam pots in which plants are placed at the four corners of each pot; the system is primarily designed for the growing of strawberry, lettuce, and herbs (Figure 15). The placement and flow of nutrient solution are similar to the vertical bag system.

An advantage of these vertical systems is the utilization of vertical space, thereby conserving lateral space if plants are grown in an enclosed shelter or greenhouse. The bag or column of pots can be rotated slowly to obtain more uniform light exposure for the plants.

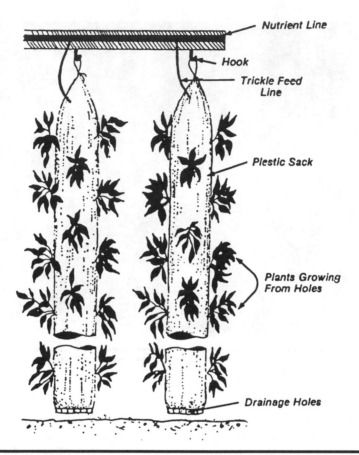

Figure 14 Hanging vertical perlite bag system for hydroponic strawberry production. (Source: Resh, 1995.)

Rockwool Slab Drip Nutrient Solution System

Rockwool is probably the most extensively used hydroponic growing material in the world today for the production of tomato, cucumber, and pepper (Bij, 1990; Ryall, 1993), although efforts are being made to find an adequate substitute because disposal of used slabs is becoming a major problem. Rockwool has excellent water-holding capacity, is relatively inert, and has proven to be an excellent substrate for plant growth (Sonneveld, 1989).

Rockwool is an inert fibrous material produced from a mixture of volcanic rock, limestone, and coke; melted at 1500 to 2000°C; extruded as fine fibers; and pressed into loosely woven sheets. The sheets are made into

Figure 15   Vertical pot system for hydroponic production (Verti-Gro, Inc., 1556 Yardley Court, Kissimmee, FL 34744).

slabs of varying widths (6 to 18 inches), normally 36 inches in length, and ranging in depth from 2 to 3 inches. The slabs are normally wrapped with white polyethylene sheets, as illustrated in Figure 16.

The slabs are normally laid flat on a prepared floor surface, which is usually first covered by white polyethylene ground sheeting. Spacing among the slabs will depend on the configuration of the growing area and the crop to be grown. Once the slabs are set in place, cuts are made along the lower edge of each slab of the polyethylene slab covering on the bottom to allow excess water to flow from the slab. An access hole is then cut on the top of the slab sheeting; it should accommodate a rockwool block containing a growing plant which will be set on the access hole. Nutrient solution is then delivered to each rockwool cube by means of a drip irrigation system. A visual sketch of a slab with a plant-containing cube and the drip tube in place is shown in Figure 17.

Today, much of the decision making as to the formulation and applica-

Figure 16   Typical rockwool slab with cucumber plants set in rockwool cube. (Courtesy of Grodan®.)

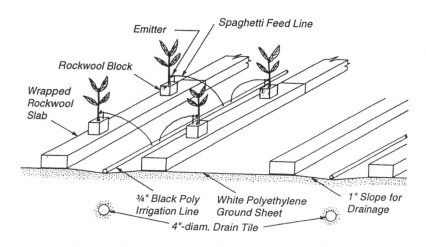

Figure 17   Illustration of the placement and arrangement for a typical rockwool slab hydroponic system. (Source: Resh, 1995.)

tion schedules for the nutrient solution is done by computer models with inputs from environmental (mainly temperature, light, etc.) and plant measurements (i.e., stage of growth, etc.). Nutrient solution formulas are similar to the solution used for NFT growing systems, with variations made based on crop and stage of growth (see Table 37; Hochmuth, 1996). Being an open system, the nutrient solution is not recovered, and that delivered is sufficient for an excess flow from the cut openings on the bottom edge of the slab. Periodically, a solution sample is drawn from the slab, its EC is determined, and the slab is leached with water if the EC is above a certain level. A pH measurement may also be made, and the nutrient solution composition may be changed if required. Normally, the elemental content of the slab-retained nutrient solution is not determined, although Ingratta et al. (1985) have given optimum and acceptable ranges for the solution of two crops, tomato and cucumber; the values are given in Table 38. These same values would also apply to other inert substrates, such as perlite.

Table 38 Optimum concentrations and acceptable ranges of the nutrient solution in the substrate

	Tomato		Cucumber	
Determination	*Optimum*	*Acceptable Range*	*Optimum*	*Acceptable Range*
EC (mS/cm)	2.5	2.0–3.0	2.0	1.5–2.5
pH	5.5	5.0–6.0	5.5	5–6
		mg/L (ppm)		
Bicarbonate (HCO_3)	<60	0–60	60	0–60
Nitrate (NO_3)	560	370–930	620	440–800
Ammonium (NH_4)	<10	0–10	<10	1–10
Phosphorus (P)	30	15–45	30	15–45
Potassium (K)	200	160–270	175	140–270
Calcium (Ca)	200	160–280	200	140–280
Magnesium (Mg)	50	25–70	50	25–70
Sulfate (SO_4)	200	100–500	200	50–300
Boron (B)	0.4	0.2–0.8	0.4	0.2–0.8
Copper (Cu)	0.04	0.02–0.1	0.04	0.02–0.1
Iron (Fe)	0.8	0.4–1.1	0.7	0.4–1.1
Manganese (Mn)	0.4	0.2–0.8	0.4	0.2–0.8
Zinc (Zn)	0.3	0.2–0.7	0.3	0.2–0.7

Source: Ingratta et al., 1985.

Although rockwool culture is widely practiced today and is the most commonly used growth medium for the hydroponic production of vegetables (tomato, cucumber, and pepper), disposal of the used slabs is a major environmental problem. Today, methods of reconstituting the rockwool slabs are being studied, with the hope of solving this disposal problem through recycling.

Organic Media Soilless Culture 10

Sphagnum peat moss and pine bark are the primary ingredients in most organic soilless cultures. Commercially prepared organic soilless mixes are readily available which have been designed for a particular use and/or crop; the mix characteristics are usually set by the manufacturer. These organic rooting media have the advantages of low cost and ease of use. It is common practice to add other materials, including vermiculite, perlite, and sand, to the organic substrate to provide desired characteristics, such as increased porosity, water-holding capacity, or weight. Although much has been said and written about the constitution of a desirable mix (Bunt, 1988), few data to substantiate these claims are generally available. Therefore, most growers depend upon past experience when selecting a commercially made mix or when making their own.

Physical and Chemical Properties

Organic media have physical and chemical properties that make their use unique compared to inorganic media. For example, sphagnum peat moss (Bunt, 1988) and pine bark (Pokorny, 1979; Ogden et al., 1987) exhibit to some degree both adsorptive and absorptive properties (Bruce et al., 1980) and thus act more like soil; these characteristics are not found in the inorganic substances, such as gravel, sand, perlite, and rockwool. These organic substances provide a *buffering capacity* which can work to the advantage of the grower, serving as a storage mechanism for the essential elements, which reduces the likelihood of both elemental excesses and shortages. In addition, the organic substances used may intrinsically contain some of the essential elements required by plants in sufficient quantity to satisfy the crop requirement.

Many organic soilless mixes are various combinations of sphagnum peat moss, pine bark, and vermiculite. In some instances, the composition of the mix may reflect the cost and availability of the major ingredient materials more than the physical and chemical characteristics they give to the mix. For example, the increased cost and reduced availability of sphagnum peat moss have led to substitution of other materials, such as pine bark (Pokorny, 1979).

Composts of various kinds, such as coarse sawdust, composted garbage and other organic refuse, and sewage sludges, have been added to mixes (Carlile and Sweetland, 1983; Handreck and Black, 1993). Their relatively low cost and the need for disposal have led to the introduction of these composted materials into some organic soilless mixes. Unfortunately, some composts contain heavy metal residues which, if present in high concentrations, are toxic to plants. Cadmium, chromium, copper, lead, manganese, and zinc are common elements found in garbage and sewage composts (Chaney, 1983). While these composts can be treated to reduce heavy metal concentrations to below levels toxic to plants, their use should be limited to the growing of non-edible crops.

Particle size and distribution in a soilless organic mix are important, as they determine both the water-holding capacity and aeration of the mix. High water-holding capacity and humid air spaces in the mix are important for germination and seedling and cutting growth, while good aeration and moderate water-holding capacity are essential for long-term plantings. A fine-particle mix is best for seed germination and short-term plant production of seedlings and cuttings; the coarse mixes are best for long-term use, such as growing potted flowering and woody ornamental plants. In a fine mix, the majority of the particles are less than 0.59 mm in diameter and will pass a NBS sieve number 20. The majority of the particles in a course mix will not pass a NBS sieve number 8, as they are 2.38 mm or larger in diameter.

For long-term container growing, the percentage of particles less than 0.59 mm in size should not exceed 20 to 30% to minimize water-logging. On the other hand, for short-term use, coarse particles 2.38 mm or larger should be completely removed from the mix.

In general, mixtures of sphagnum peat moss and vermiculite and/or perlite are the majority ingredients in fine-particle organic mixtures, while pine bark alone or with perlite constitutes most of the coarser mixes. However, this is a generalization that does not entirely hold for pine bark, which can be processed to make a fine-particle mix similar to sphagnum–peat-moss-based mixes.

A common component added to some mixes is sand; it is added to provide porosity in fine mixes or weight when needed to keep plant containers upright in either fine or coarse mixes. However, sand should not constitute more than 20 to 25% of the mix. If more than 50% of a mix is sand, weight and reduced water-holding capacity become a problem. The recommended grade of sand is "builder's" sand, which is a coarse-particle sand; 100% passes a 10-mesh sieve but only 30% a 40-mesh sieve.

Segregation will occur during the preparation and handling of an organic soilless mix, as most of the components (sphagnum peat moss, pine bark, perlite, vermiculite, sand, limestone, fertilizer, etc.) vary in particle size and density. Therefore, care is required when preparing, mixing, and handling to prevent segregation. This is particularly important when fertilizer ingredients are being blended into a mix. Even when using an automatic potting machine, an organic soilless mix may be segregated as it is moved from the mixing bin to the pot-filling chute. Making or keeping the mix slightly moist during handling and mixing helps keep components from segregating easily.

Segregation of components is a common problem in prepared mixes due to separation during shipment, as the less dense and larger particles move upward through the mix. Upon receipt and prior to use, careful turning of the mix may be required to restore the materials to their original blend.

Organic Soilless Mix Formulas

While there are many formulas for preparing organic soilless mixes today, the basic concepts of formulation were established by Sheldrake and Boodley (1965) and Boodley and Sheldrake (1972) in their Cornell Peat-Lite mixes and from the University of California basic mix (Baker, 1957); the formulations are given in Tables 39 and 40, respectively. From these basic formulas, a number of related mixes for various uses have been devised. A list of ingredients for some of these mixes is given in Table 41.

Table 39   Instructions for three Peat-Lite mixes

Ingredients	Amount
Peat-Lite Mix A	
Sphagnum peat moss	11 bu.
Horticultural vermiculite No. 2	11 bu.
Limestone, ground	5 lb.
Superphosphate 0-20-0	1 lb.
5-10-5 fertilizer	2 to 12 lb.
Peat-Lite Mix B	
Sphagnum peat moss	11 bu.
Horticultural perlite	11 bu.
Limestone, ground	5 lb.
Superphosphate, 0-20-0	1 lb.
5-10-5 fertilizer	2 to 12 lb.
Peat-Lite Mix C (for germinating seed)	
Sphagnum peat moss	1 bu.
Horticultural perlite	1 bu.
Limestone, dolomitic	7.5 oz.
Superphosphate, 0-20-0	1.5 oz.
Ammonium nitrate	1 oz.

Source: Sheldrake and Boodley, 1965; Boodley and Sheldrake, 1972.

Long-term use of an organic soilless mix requires a different nutrient element charge, as illustrated by three different types of mixes and use: a pine-bark mix for container-grown nursery stock (Table 42), a growing-on mix using sphagnum peat moss for ornamental plant production (Table 43), and a mix for the bag culture of tomato (Table 44).

The Tapia (1985) mix using pine bark is used for growing seedlings, vegetables, and nursery plants and is designed for use with low and high calcium-containing irrigation water; the formulas are given in Table 45.

At this point, the reader may be confused as to the proper constitution and use of an organic soilless mix, and indeed this confusion is widespread. There are no set rules. There is little commercial literature on the use of organic soilless mixes for plant production and little uniformity as to technique of growing. For example, in the *Ball Red Book* (Ball, 1985), a widely used 720-page text on the greenhouse culture of plants, only 3 pages are devoted to descriptions of soilless mixes and their use.

Table 40 Basic fertilizer additions to make the University of California mix (50% sand and 50% peat moss to make one cubic yard)

Added Ingredients	Amount
Hoof-and-horn or blood meal (13%)	2.5 lb.
Potassium nitrate	4.0 oz.
Potassium sulfate	4.0 oz.
Superphosphate, 0-20-0	2.5 lb.
Dolomite limestone	7.5 lb.
Calcitic limestone	2.5 lb.

Source: Baker, 1957.

Table 41 Ingredients to make one cubic yard of soilless organic mix

		Type of Mix				
Ingredients	Cornell Peat-Lite*	U.C. Mix #D*	U.C. Mix #E*	Canada Mix Seedling*	NJ Tomato Greenhouse	Georgia Greenhouse Tomato
Sphagnum peat moss	11 bu.	16.5 bu.	22 bu.	12 bu.	9 bu.	—
Milled pine bark	—	—	—	—	—	9 bu.
Vermiculite	11 bu.	—	—	10 bu.	9 bu.	—
Perlite	—	—	—	—	4 bu.	—
Sand	—	5.5 bu.	—	—	—	—
Limestone	5 lb.	9 lb.	7.5 lb.	4 lb.	8 lb.	1 lb.
Superphosphate, 0-20-0	2 lb.	2 lb.	1 lb.	1 lb.	2 lb.	—
5-10-5 fertilizer	6 lb.	—	—	—	—	—
10-10-10 fertilizer	—	—	—	2 lb.	—	1 lb.
Potassium nitrate	—	—	0.3 lb.	0.5 lb.	—	—
Calcium nitrate	—	—	—	—	1 lb.	—
Borax	10 g.	—	—	1 g.	10 g.	—
Chelated iron	25 g.	—	—	—	35 g.	—

* These mixes are mainly for short-term growth, whereas the other mixes are for long-term greenhouse tomato production.

It is evident that for short-term growing, a wide range of conditions can be tolerated in terms of mix constitutes and methods of fertilization. The grower's observation and experience are the primary controls. By adding or

Table 42   Pine bark mix for container-grown nursery stock developed at the Levin Horticultural Research Centre, New Zealand*

Added Ingredients	kg/m^3	yd^3
Superphosphate	1.0	1 lb. 11 oz.
Calcium ammonium nitrate	1.0	1 lb. 11 oz.
Osmocote, 18-11-10 (9 month)	3.0	5 lb.
Dolomitic limestone	4–5	6–8.5 lb.
Trace element mixture**		

* Pine bark particle size distribution: 100% <5 mm, 70–85% <2.5 mm, 30–60% <1 mm, and 10–20% <0.5 mm.

** Borax = 11.8 g/m^3; copper sulfate = 35.4 g/m^3; ferrous sulfate = 50 g/m^3; chelated iron = 14.2 g/m^3; manganese sulfate = 14.2 g/m^3; zinc sulfate = 2.4 g/m^3.

Source: Bunt, 1988.

Table 43   Sphagnum peat moss mix for ornamental plant production

Added Ingredients	kg/m^3	yd^3
Superphosphate, 0-20-0	0.9	1 lb. 8 oz.
Potassium nitrate	0.3	8 oz.
20-19-18 fertilizer	0.15	4 oz.
Dolomitic limestone	3.0	5 lb.

Source: White, 1974.

Table 44   Peat nodules (sedge or humified sphagnum) in 20-L bag for greenhouse tomato culture

Added Ingredients	kg/m^3	yd^3
Superphosphate, 0-20-0	1.75	3 lb.
Potassium nitrate	0.87	1 lb. 8 oz.
Potassium sulfate	0.44	12 oz.
Ground limestone	4.2	7 lb.
Dolomitic limestone	3.0	5 lb.
Frit 253A	0.4	10 oz.

Note: Additional slow-release nitrogen as 0.44 kg/m^3 urea-formaldehyde (167 mg N/L) is sometimes included. If a slow-release phosphorus fertilizer is required, magnesium ammonium phosphate ("MagAmp" or "Enmag") at 1.5 kg/m^3 is added.

Source: Bunt, 1988.

Table 45   Elements and liming materials added to pine bark for growing seedlings, vegetables, and nursery plants using two calcium regimes

Added Elements	*g/cm³*	
	General	*Mixture Low Calcium**
Nitrogen (N)	450	450
Phosphorus (P)	150	150
Potassium (K)	200	200
Calcium (Ca)	2300	—
Magnesium (Mg)	650	—
FRIT	300	300
Liming material		
Dolomitic lime	4000	—
Calcitic lime	4000	—

* Mix for use where calcium will be supplied by the irrigation water.

Source: Tapia, 1985.

withholding fertilizer, the rate of growth or plant appearance can be easily changed. It is when an organic soilless mix is used for long-term growing that mix constitutes and fertilization technique become critical. Many of the problems that occur in other forms of soilless and hydroponic growing appear, such as soluble salt accumulation, disease control, pH shifts, and nutrient element stress. Only by observation and testing can the grower control these factors in order to prevent reduction in plant growth and yield.

Limestone and Fertilizer Additions

Limestone has traditionally been added to organic soilless mixes to both raise the water pH of the mix and provide a source of both calcium and magnesium, which are essential elements. However, recent research raises questions about this practice, as raising the pH of the mix even moderately can significantly reduce the availability of most of the essential elements, as shown in Figure 18.

An organic soilless mix should not exceed a water pH of 5.5, with the optimum pH range between 4.5 and 5.5. Elimination of limestone from

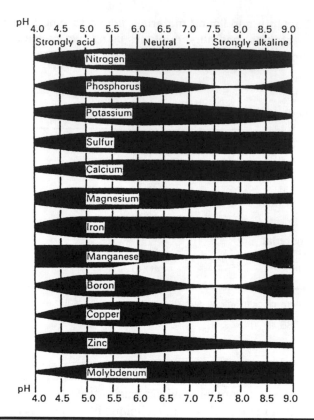

Figure 18   The effect of pH on the availability of the essential elements in an organic soilless mix. (Source: Peterson, 1981.)

organic soilless mix formulas requires a substitute source for calcium either as calcium sulfate ($CaSO_4$) or calcium nitrate [$Ca(NO_3)_2 \cdot 4H_2O$] and for magnesium as magnesium sulfate ($MgSO_4 \cdot 7H_2O$).

In addition, the quality of irrigation water can alter, with time, the pH of an organic soilless mix if the water contains sizable quantities (>30 mg/L, ppm) of either calcium or magnesium, or both. With each irrigation, the mix is essentially "limed," and the water pH of the mix rises a bit. In time, the pH may reach the point where the availability of some of the essential elements is adversely affected and one or more nutrient element deficiencies occur. The problem can be partially solved by not adding calcium- or magnesium-containing sources to the mix initially, thereby relying on the calcium and magnesium content of the irrigation water to supply these two essential plant elements. In such a case, it is essential for the grower to know if calcium or magnesium has been added to the mix or if the mix is for use where the calcium and magnesium requirements of the plant are to be supplied by the irrigation water.

The addition of nitrogen, phosphorus, and potassium, usually as chemical fertilizer, is primarily determined based on the use of the organic soilless mix. From a practical standpoint and for short-term cropping (growth period of less than 8 weeks), these elements and the other essential elements would be added to the mix when constituted. However, from a control standpoint, and for all long-term cropping, adding the essential elements as required is best. For practical reasons, some of the elements, such as the micronutrients, may be added when constituting the mix, reserving the three major elements nitrogen, phosphorus, and potassium for addition as required by the crop being grown. Unfortunately, there is no one best way that can be recommended. The best compromise between considerations of practicality and control appears to be adding the micronutrients and the major elements phosphorus, potassium, calcium, and magnesium to the organic soilless mix initially and then supplementing as required based either on a plant analysis and an assay of the mix or on plant growth and appearance; the major elements, primarily nitrogen and potassium, can be added periodically to satisfy the crop requirement based on growth and plant appearance. The element phosphorus may be added to the latter group if a complete nitrogen-phosphorus-potassium fertilizer is used to supply the required nitrogen and potassium.

Liquid fertilizers, such as 20-20-20 (N-P_2O-K_2O), are frequently used for supplementation by addition to the irrigation water. The concentration is varied depending on the crop requirement. A common recommendation is that the concentration of nitrogen be between 50 and 100 mg/L (ppm). A list of materials and the amount required to prepare fertilizer solutions with a nitrogen concentration of 50, 100, 150, and 200 mg/L (ppm) are given in Table 46, and ingredient concentrations to make a series of nitrogen-phosphorus-potassium-containing solutions are given in Tables 47 and 48.

Growers should be aware that repeated long-term use of a fertilizer, such as 20-20-20 applied through the irrigation water, can lead to excesses in phosphorus if this element has already been added to the mix. Therefore, care should be taken to ensure that phosphorus excess does not occur either by not putting it into the mix initially or by selecting a liquid fertilizer that does not contain phosphorus.

Table 46   Pounds of fertilizer per 100 gallons of water to make solutions with nitrogen concentration of 50, 100, 150, and 200 mg/L (ppm) for fertilizer supplementation

Fertilizer	Milligrams Nitrogen per Liter (ppm)			
	50	100	150	200
	Pounds per 100 Gallons Water			
Calcium nitrate [Ca(NO$_3$)$_2$·4H$_2$O]	0.24	0.48	0.72	0.96
Potassium nitrate (KNO$_3$)	0.32	0.64	0.96	1.28
5-10-5	0.83	1.66	2.49	3.32
10-10-10	0.41	0.83	1.29	1.66
20-20-20	0.20	0.41	0.63	0.83

Slow-release fertilizers are added to an organic soilless mix for elemental release control. Osmocote (Grace-Sierra Horticultural Products Co., 1001 Yosemite Drive, Mippitas, CA 95035), MagAmp, ureaform, and ordinary chemical-based fertilizer in small perforated polyethylene bags placed in the mix give some degree of control by providing a steady supply of essential elements to plants during their growth cycle, as well as reducing leaching losses. Osmocote, for example, can be obtained in various formulations with varying release-rate characteristics. However, the high cost of some of these slow-release fertilizers must be balanced against the advantages of the control obtained.

Growing Techniques

Traditional organic soilless media culture is carried out with the medium placed in a bed, pot or can. Water, with or without fertilizer added, is applied periodically by either overhead irrigation or by drip irrigation into the container in quantities relative to the atmospheric demand on the plant. As is the case in gravel and sand systems, the medium may require periodic flushing with water to remove accumulated salts; the need to flush is determined by a soluble salt reading of the medium itself or the effluent from the container. Commonly, the container is discarded after one use, although some growers have devised interesting schemes to use the medium for more

Table 47 Weight of fertilizer (in grams) required to prepare one liter of stock solution for dilution at 1 in 200 to give a range of liquid feeds

Fertilizer*	g/L						
	N50 P7.5	N100 P15	N150 P22.5	N200 P30	N250 P37.5	N300 P45	
Ammonium nitrate	16.0	42.3	68.9	95.4	121.8	148.3	
Monoammonium phosphate	6.1	12.3	18.4	24.5	30.6	36.7	K50
Potassium nitrate	26.3	26.3	26.3	26.3	26.3	26.3	
(EC)	(0.32)	(0.57)	(0.80)	(1.04)	(1.28)	(1.51)	
Ammonium nitrate	5.4	31.9	58.4	84.8	111.3	137.8	
Monoammonium phosphate	6.1	12.3	18.4	24.5	30.6	36.7	K100
Potassium nitrate	52.6	52.6	52.6	52.6	52.6	52.6	
(EC)	(0.41)	(0.65)	(0.89)	(1.12)	(1.35)	(1.58)	
Ammonium nitrate	**	21.4	47.8	74.3	100.8	127.3	
Monoammonium phosphate	6.1	12.3	18.4	24.5	30.6	36.7	K150
Potassium nitrate	78.9	78.9	78.9	78.9	78.9	78.9	
(EC)	(0.58)	(0.73)	(0.96)	(1.20)	(1.43)	(1.66)	
Ammonium nitrate	—	10.8	37.3	63.8	90.2	116.7	
Monoammonium phosphate	—	12.3	18.4	24.5	30.6	36.7	K200
Potassium nitrate	—	105.3	105.3	105.3	105.3	105.3	
(EC)	—	(0.82)	(1.04)	(1.29)	(1.52)	(1.74)	
Ammonium nitrate	—	nil	26.8	53.3	79.8	106.3	
Monoammonium phosphate	—	12.3	18.4	24.5	30.6	36.7	K250
Potassium nitrate	—	131.5	131.5	131.5	131.5	131.5	
(EC)	—	(0.90)	(1.11)	(1.37)	(1.60)	(1.84)	
Ammonium nitrate	—	***	16.3	42.8	69.2	95.7	
Monoammonium phosphate	—	12.3	18.4	24.5	30.6	36.7	K300
Potassium nitrate	—	157.8	157.8	157.8	157.8	157.8	
(EC)	—	—	(1.22)	(1.44)	(1.69)	(1.87)	

Note: Numbers following N, P, and K are mg/L (ppm) of nitrogen, phosphorus, and potassium in the prepared solutions.

* This table was computed using the following fertilizer analysis: ammonium nitrate = 35% N; monoammonium phosphate = 12% N, 24.5% P; potassium nitrate = 14% N, 38% K. The EC values of the diluted fertilizer solutions are expressed in mmhos/cm at 25°C.

** 58.9 mg/L (ppm) nitrogen.

*** 117.8 mg/L (ppm) nitrogen.

Source: Bunt, 1988.

Table 48   Preparation of liquid feeds from complete nitrogen-phosphorus-potassium water-soluble fertilizer (based on U.S. gallon)

Fertilizer			Element Equivalent (%)			Ounces per U.S. Gallon Diluted 1 in 200	mg/L (ppm)		
N	P_2O_5	K_2O	N	P	K		N	P	K
20	20	20	20	8.8	16.6	26.6	200	88	166
15	30	15	15	13.2	12.5	35.5	200	176	167
14	14	14	14	6.2	11.6	38.0	200	88	166
21	7	7	21	3.1	5.8	25.3	200	30	55
20	5	30	20	2.2	24.9	26.6	200	22	249
25	10	10	25	4.4	8.3	21.3	200	35	66

Note: To obtain 100 mg N/L (ppm), use one-half the above weights of fertilizer; for 150 mg N/L (ppm), use three-quarters of the weight.

than one crop; one example is growing a crop of tomato followed by an outdoor ornamental tree or shrub. The sale of the ornamental plant also provides a means of disposal of the container and medium.

Another growing technique is to plant directly into the medium shipping bag, adding the required essential elements and water by drip irrigation. Normally, the mix is supplemented with micronutrients, and the major elements are applied in a nutrient solution whose composition may be comparable to that of a Hoagland nutrient solution (see Table 11), with or without micronutrients, or some other formulation. An example of nutrient solutions recommended for pre-enriched pine bark and sawdust for cucumber and tomato is given in Table 49.

The flow of the nutrient solution through the drip irrigation system must be sufficient to meet the water requirement. If plant growth is normal, elemental utilization should be sufficient to prevent a significant accumulation of excess salts. When plant growth is slow due to poor external growing conditions, then applying only water without elements added is best, and the nutrient solution application is resumed when growth conditions improve. Some growers substitute a mixture of an equal ratio of potassium and calcium nitrate [KNO_3 and $Ca(NO_3)_2 \cdot 4H_2O$, respectively] to give a solution containing 100 mg/L (ppm) nitrogen in place of the Hoagland formula. If this is done, the medium must contain sufficient phosphorus and magnesium to meet the crop requirement.

Table 49   Nutrient solutions recommended for plants in pre-enriched bark

| Element | mg/L (ppm) | | | |
| | Pre-Enriched Bark | | Sawdust | |
	Seedlings Potted Plants Cucumber to 1st Harvest	Cucumber After 1st Harvest	Tomato	Cucumber
Major elements				
Nitrogen (N)	200	260	170	185
Phosphorus (P)	25	50	55	40
Potassium (K)	150	150	400	210
Calcium (Ca)	—	330	160	210
Magnesium (Mg)	—	50	50	25
Micronutrients				
Iron (Fe)	—	2.5	1.2	1.0
Manganese (Mn)	—	0.62	1.0	0.3
Boron (B)	—	0.44	0.5	0.7
Copper (Cu)	—	0.05	0.03	0.03
Zinc (Zn)	—	0.03	0.11	0.1
Molybdenum (Mo)	—	0.09	0.02	0.05

Source: Smith, 1985.

A Unique Application

A unique application of an organic soilless mix for long-term growing is a system employing subirrigation based on a technique first introduced by Geraldson (1963) for growing staked tomatoes in the sandy soils of south-western Florida (Geraldson, 1982). In fields where the water table level can be controlled, raised plastic covered beds are prepared with a band of fertilizer placed down each side of the bed. Tomato plants are set in the center of the beds, and the roots grow into the area of the soil which is balanced in water and elemental content, as illustrated in Figure 19. The author duplicated the same system by placing a coarse organic soilless mix into a watertight box so that a constant water table could be maintained under the mix.

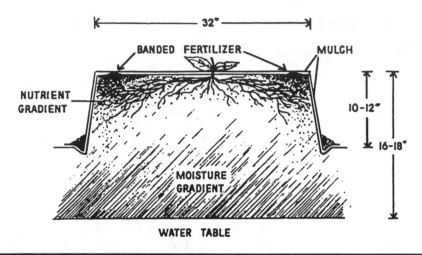

Figure 19 **Control of the root ionic environment obtained by banding fertilizer on the surface of a raised bed, using a plastic mulch cover and maintaining a definite water table. (Source: Geraldson, 1963.)**

A 4′ × 30′ (1 × 9 m) grow-box system, employing pine bark as the growing medium supplemented with limestone and fertilizer (as shown in the last column of Table 41), was successfully used to grow greenhouse tomato, cucumber, and snapdragons over an 8-year period (Figure 20). The constant water table is maintained under 7 inches (18 cm) of pine bark with an automatic float value system. No water is applied overhead. Fertilizer is added to the medium between crops based on an elemental assay of the medium. The tomato crop is supplemented periodically with a mixture of potassium and calcium nitrate [KNO_3 and $Ca(NO_3)_2 \cdot 4H_2O$, respectively] at a nitrogen concentration of 100 mg/L (ppm) based on a need determined by the physical appearance of the plant and/or periodic plant analyses.

The pine bark maintained its original physical properties over the 8-year period of use, requiring only small yearly additions of new pine bark to maintain the initial volume. Initially, some boxes were filled with a Peat-Lite mix, which failed to maintain a good physical character beyond the first year and therefore was discarded and replaced by pine bark.

The success of this subirrigation grow-box system is due in large part to the constantly maintained water table, which allows the roots to grow into that portion of the medium *ideal* in terms of water, aeration, and elemental supply. The grower does not need to be concerned about watering on high

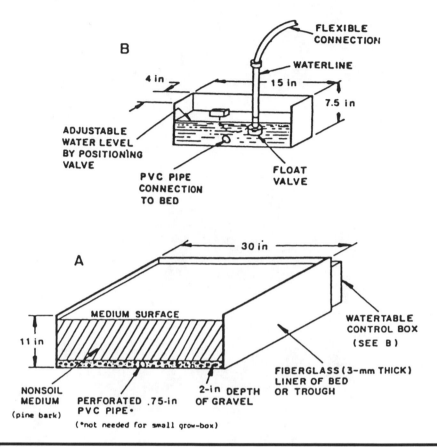

Figure 20   Pine bark grow-box system employing a constant water table with automatic water level control. (Source: Jones, 1980.)

atmospheric demand days, as water is always available to the plant. As the growing system is essentially self-regulating, the grower can concentrate on the cultural management of the crop.

The grow-box can be made in almost any size to accommodate a wide range of uses, even outdoor family vegetable gardening (Jones, 1980). The only critical dimension is the depth of the organic soilless medium, which must not be more or less than 7 inches above the water table. Pine bark seems to be the best of all the organic substances for this application.

Diagnostic Testing Procedures 11

Success with any growing system is based to a considerable degree on the ability of the grower to effectively evaluate and diagnose the condition of the crop at all times (Eysinga and Smilde, 1981; Paterson and Hall, 1981; Jones, 1993a). This is particularly true for the hydroponic/soilless culture grower and absolutely essential for the hydroponic grower, since most of the essential elements (except for carbon, hydrogen, and oxygen) required by the plant are being supplied by means of a nutrient solution. Errors in preparing and using the nutrient solution will affect plant growth, sometimes within a matter of a few days. Some growers possess a unique ability to sense when things are not right and take the proper corrective steps before significant crop damage is done. Most, however, must rely on more obvious and objective measures to assist them in determining how their growing system is working and how plants are responding to their management inputs. In the latter case, there is no substitute for systematic observation and testing. As the genetic potential of a crop is approached, every management decision becomes increasingly important. Small errors can have significant impact; therefore, there is a need to do every task without error in timing or process. Under such conditions, nutritional management becomes absolutely essential.

Laboratory testing and diagnostic services are readily available in the United States and Canada (Anon., 1992). Once the laboratory selection has been made, it is important to obtain from the laboratory its instructions for collecting and shipping samples before sending them.

With the analytical advances that have occurred in recent years, together with the ease of quickly transporting samples and analysis results, growers can almost monitor their plant growing systems on a real-time basis. Although a routine of periodic testing is time consuming and costly, the application of the results obtained can more than cover the costs in terms of a saved crop and superior quality production. The grower should get into the habit of routinely analyzing the water source, nutrient solution, growing media, and crop. Interpretations and recommendations based on assay results are designed to assist the grower in order to avoid crop losses and product quality reductions.

Water Analysis

Water available for making a nutrient solution or for irrigation may not be of sufficient quality (i.e., free from inorganic as well as organic substances) to be suitable for use. Pure water is not essential, but the degree of *impurity* needs to be determined. Even domestic water supplies, although safe for drinking, may not be suitable for plant use. Water from surface groundwater sources, ponds, lakes, and rivers is particularly suspect, while collected rainwater and deep-well water are less so.

For the elements, the presence of calcium and magnesium could be considered complementary because both elements are essential, whereas the presence of carbonate, bicarbonate, boron, sodium, chloride, fluoride, and sulfide could be considered undesirable if levels are relatively high. The maximum concentrations of these elements and ions in irrigation water and water for making a nutrient solution have been established, as presented in Chapter 7.

The only way to determine what is in the water is to have it tested for all these elements and ions. Testing for the presence of organic constituents is a decision that is based on expected presence. Surface waters may contain disease organisms and algae, while in agricultural areas, various residues from the use of herbicides and/or pesticides may be in the water. Tomato, for example, is quite sensitive to many types of organic chemicals; therefore, their presence in water could make its use undesirable.

Knowing what is in the water will determine whether it is acceptable with or without treatment and whether adjustments would be required to compensate for constituents that are present.

Nutrient Solution Analysis

Errors in the preparation of a nutrient solution are not uncommon; hence the requirement for an analysis to check on the final elemental concentrations prior to use. Since the elemental composition of the nutrient solution can be altered considerably in closed recirculating systems, it is equally important to monitor the composition of the solution as frequently as practical. A record of the analysis results should be kept and a track developed to determine how the concentration of each element changes with each passage through the rooting media. On the basis of such analyses, change schedules, replenishment needs, and crop utilization patterns can be determined. The track establishes what adjustments in the composition of the nutrient solution are needed to compensate for the "crop effect" not only for the current crop but for future crops as well.

In addition, periodic analysis allows the grower to properly supplement the nutrient solution in order to maintain consistent elemental levels to ensure good crop growth as well as extend the useful life of the nutrient solution. Significant economies can be gained by extending the life of the nutrient solution in terms of both water and chemical use.

Laboratory analysis is recommended, although on-site analysis is possible with the use of kits and relatively simple analytical devices (Schippers, 1991). It is now possible to continuously monitor the nutrient solution with devices such as specific ion and pH electrodes, conductivity meters, etc. (Raper, 1987). The grower needs to determine how best to monitor the nutrient solution based on cost and the requirements of the selected growing system.

Electrical conductivity (EC) is frequently used as a means of determining elemental replenishment needs in closed recirculating nutrient solution growing systems. This technique is useful if previous knowledge is available as to which elements are likely to change and by how much. It is far more desirable to do an analysis which quantifies each individual element and its ratio in the nutrient solution so that specific adjustments can be made to bring the nutrient solution back to its original composition.

The analysis of the nutrient solution should include pH and tests to determine the concentration of the major elements nitrogen (i.e., nitrate [NO_3^-] and ammonium [NH_4^+]), phosphorus, potassium, calcium, and magnesium). These determinations can be made on-site by using a water analysis kit (HACH Chemical Company, P.O. Box 389, Loveland, CO 80539).

Although test kit procedures are available for determination of some of the micronutrients, laboratory analysis is recommended. However, concentration monitoring of the micronutrients is not as critical as monitoring the major elements, unless a micronutrient problem is suspected. For any diagnostic problem, laboratory analysis is always recommended, including all the essential elements—both the major elements and micronutrients.

Elemental Analysis of the Growth Medium

Elemental analysis of plant growth medium is an important part of the total evaluation of the elemental status of the media-crop system. When coupled with a plant analysis, it allows the grower to determine what elemental stresses exist and how best to bring them under control. This analysis may be comprehensive, to determine the concentration present in the growth medium by element, or more general, measuring the total soluble salt content or effluent from the medium or by extraction of an equilibrium solution. A comprehensive test is more valuable as a means of pinpointing possible elemental problems than just a determination of the EC of the effluent or an extracted solution.

A test of an inorganic growth medium, such as gravel, sand, perlite, and rockwool, measures the accumulation of salts that will significantly affect the elemental composition of the nutrient solution being circulated through it. Knowing what is accumulating in the growth medium, it then becomes possible to alter the nutrient solution composition sufficiently to utilize the accumulated elements or to begin to make adjustments in the nutrient solution formula with the idea of reducing the rate of accumulation while partially utilizing those elements already part of the medium.

For those using perlite bags or rockwool slabs, the recommendation today is to periodically draw solution with a syringe from the bag or slab for assay. Based on either a complete analysis of this solution or only its EC, leaching may be recommended to remove accumulated salts. In some management schemes, leaching of the growth medium is done on a regular schedule as a matter of normal routine. Systems using regularly scheduled leaching should also be subjected to periodic analysis of the growth medium effluent to confirm that the leaching schedule is in fact doing the job intended.

For an organic growth medium, such as peat mixtures, pine bark, etc., the sampling and assay procedures are quite different. Monitoring of the

Table 50 General information guidelines for organic rooting media analyzed by the saturated extract method

Test Parameter	Category (mg/kg, ppm)				
	Low	Acceptable	Optimum	High	Very High
Nitrate-N	0–39	40–99	100–199	200–299	300+
Phosphorus (P)	0–2	3–5	6–9	11–18	19+
Potassium (K)	0–59	60–149	150–249	250–349	350+
Calcium (Ca)	0–79	80–199	200+	—	—
Magnesium (Mg)	0–29	30–69	70+	—	—
Soluble salts (mS/cm)	<0.75	0.75–2.0	2.0–3.5	3.5–5.0	5.0+

Source: Warnke, 1988.

medium is not necessary as a matter of routine, but a test should be made at its initial use, whenever plant stress appears, or when a significant change in a cultural practice occurs. Cores of media taken to the rooting depth or to the bottom of the growing vessel are randomly collected and composited, and the composite is sent to the laboratory for analysis. The ranges in concentration for the various elements, pH, and soluble salts have been established by researchers at Michigan State University (Warnke, 1986) as shown in Table 50.

Although the testing procedures are quite different for each growing medium, the objective of the analysis is the same—to evaluate the elemental status of the medium for diagnostic purposes. The elements present in the growth medium serve as a major contributor toward meeting the crop requirement. Therefore, one objective for an analysis is to determine the level of each of the essential elements in the growing medium that will contribute toward satisfying the crop requirement.

The other purpose of medium analysis is to track preferential element accumulation by the medium. In systems where the bulk of the elemental requirement is supplied by the nutrient solution, growth medium analysis serves to determine accumulation rates so as to avoid imbalances and potential toxicities. In such cases, an EC measurement of the effluent from the medium, or an extraction of it, may be sufficient.

By tracking, the elemental composition of the growth medium can be followed and adjustments made based on changing concentrations away from or beyond the sufficiency range. Therefore, these periodic analyses become the means for regulating the input of the essential elements in order to prevent deficiencies or excesses from occurring.

Plant Analysis

The objective of a plant analysis (sometimes referred to as *leaf analysis*) is to monitor the elemental content of the plant in order to ensure that all of the essential elements are being supplied in sufficient quantity to satisfy the crop requirement, as well as assuring against elemental imbalances and excesses (Berry and Wallace, 1981; Faulkner, 1993; Mills and Jones, 1996). The grower should develop a routine of sampling and analysis during critical periods in the growth cycle (Bloom, 1987). Such a system of periodic sampling and analysis based on various stages within the growth cycle has been suggested by Tapia (1985) for tomato and pepper, as shown in Tables 51 and 52, respectively.

Unfortunately, plant analysis has largely been thought of as a diagnostic device, while its usefulness for monitoring is of greater significance. The procedure of routine sampling and analysis is frequently referred to as *tracking*. Tracking provides the information needed to establish what nutrient solution management procedure is required to ensure that all of the essential element levels are within the sufficiency range for the crop being

Table 51   Tomato plant development in 12 samplings

Sampling Number	Development Stage
1	Four to five true leaves
2	Eight true leaves
3	First cluster in anthesis
4	Second cluster in anthesis and first cluster with set fruits
5	Third cluster in blossom, second cluster in anthesis, first cluster with developing fruits
6	Third cluster in anthesis, second and first clusters with developing fruits, fourth and fifth clusters in blossom
7	Fourth cluster in anthesis
8	Fifth cluster in anthesis
9–12	Full production stages

Source: Tapia, 1985.

Table 52   Sweet pepper plant development in ten samplings

Sampling Number	Development Stage
1	Four to six expanded leaves
2	Seven to nine expanded leaves
3	First flower in blossom
4	Five to thirteen flowers in blossom
5	One to thirteen flowers in anthesis
6	Full anthesis and starting set fruits
7	Full set fruits
8	Developing fruits
9–10	Full production stages

Source: Tapia, 1985.

grown. It is well worth the time and expense to develop a track of elemental sufficiency in order to firmly establish the proper nutrient solution management system for future use.

The diagnostic role for plant (leaf) analysis is equally important. A grower faced with a suspected essential element deficiency or imbalance should verify the suspected insufficiency by means of plant (leaf) analysis. Many symptoms of elemental stress are quite similar and can fool the best trained grower. In addition, most any stress can be due to the relationship between or among the elements and therefore may require more than just a simple change in the nutrient solution formula to correct it. Without an analysis result, a change could be made which would only further aggravate the problem. An excellent review on the principles and practice of plant analysis has been given by Munson and Nelson (1990). The techniques of sampling, handling, and analyzing plant tissue have been reviewed by Jones and Case (1990).

Since a plant leaf analysis requires the use of a competent laboratory, contact with the laboratory should be made before samples are collected and sent. Most laboratories have specific sampling and submission procedures which are important to follow. Sampling procedures for several commonly grown hydroponic crops are given in Table 53.

If no specific sampling procedures are given or known for a particular plant, including the time for sampling, the rule of thumb is to collect recently mature leaves below the growing point. Normally, the times for

Table 53   Recommended sampling procedures for cucumber, lettuce, pepper, and tomato

Crop	Plant Part	Time	Number of Plants to Sample
Cucumber	Fifth leaf from top	First fully developed	12
Lettuce	Wrapper leaf	Mature	12
Pepper	Recent fully developed leaf	First bloom	25
	Recent fully developed leaf	Mid-season	25
Tomato	Leaf opposite and below most recent fruit cluster	When setting fruit	12

sampling are scheduled at major changes in the growth cycle, such as at flowering, initial fruit set, etc. In addition, these same sampling procedures should be followed if the plant is being monitored periodically over the course of its life cycle; this procedure is necessary to maintain a track of the elemental content of the plant.

For diagnostic testing, when visual symptoms of plant stress are evident, it is advisable to take similar plant tissues from both *affected* and *normal* plants. In this way, a comparison of analytical results can be made, which is far more helpful in the interpretation than just an analysis of the stressed plants alone.

Great care should be used when selecting plants for sampling as well as when selecting the plant part. In addition to what should be sampled, there are also avoidance criteria as to what not to sample or include in the sample:

- Diseased, insect-damaged or mechanically damaged plants or tissues
- Dead plant tissue
- Dusty or chemical-coated tissue

Tissue that is covered with dust or chemicals can be decontaminated by careful washing using the following procedure:

- Prepare a 2% detergent solution and place in a large container.
- Place the fresh leaf tissue in the detergent solution and gently rub with the fingers for no longer than 15 seconds.
- Remove the tissue from the detergent solution and quickly rinse in a stream of flowing pure water.
- Blot dry with a clean cloth or paper towel.

Great care is needed to ensure that the tissue being "washed" is not being contaminated by some other substance present in the wash water or by contact with other substances or that the elements potassium and boron are not being lost from the tissue in the washing process, as both can be easily leached if the exposure time to the washing and rinsing water is longer than that specified in the instructions.

Once the tissues have been collected, it is best to air dry them (one day in the open air is usually sufficient) before shipping to the laboratory for analysis. This will keep them from rotting while in transit, as any loss in dry weight will affect the analysis result. Details on sampling, sample preparation, and analysis of plant tissue have been given by Jones and Case (1990).

An interpretation of an analysis result is done by comparing the assay result obtained with established critical values or sufficiency ranges; the latter is the more commonly used category for evaluation (Martin-Prével et al., 1987; Reuter and Robinson, 1986; Jones et al., 1991). Categories of sufficiency for bell pepper, cucumber, lettuce, and tomato from a number of sources are given in Tables 54 through 67. Note that these interpretative

Table 54   Elemental concentrations in upper fully developed leaf of bell pepper for diagnostic evaluation

	Concentration in Upper Fully Developed Leaf (%)		
Element	Low	Sufficient	High
Major Elements			
Nitrogen (N)	3.00–3.49	3.50–5.0	>5.0
Phosphorus (P)	0.18–0.21	0.22–0.7	>0.8
Potassium (K)	3.00–3.49	3.50–4.5	>4.5
Calcium (Ca)	1.00–1.29	1.30–2.8	>2.8
Magnesium (Mg)	0.26–0.29	0.30–1.0	>1.0
		(ppm)	
Micronutrients			
Boron (B)	23–24	25–75	>75
Copper (Cu)	4–5	6–25	>25
Iron (Fe)	50–59	60–300	>300
Manganese (Mn)	40–49	50–250	>250
Zinc (Zn)	18–19	20–200	>200

Source: Mills and Jones, 1996.

Table 55   Elemental concentrations in leaf of cucumber plant for the first fully developed leaf for diagnostic evaluation

	Concentration in First Fully Developed Leaf (%)		
Element	Low	Sufficient	High
Major Elements			
Nitrogen (N)	3.50–4.29	4.30–6.0	>6.0
Phosphorus (P)	0.25–0.29	0.30–0.7	>0.7
Potassium (K)	2.00–3.09	3.10–5.5	>5.5
Calcium (Ca)	1.50–2.49	2.50–4.0	>4.0
Magnesium (Mg)	0.25–0.34	0.35–1.0	>1.0
Sulfur (S)	<0.40	0.40–0.7	>0.7
		(ppm)	
Micronutrients			
Boron (B)	25–29	30–100	>100
Copper (Cu)	6–7	8–10	>10
Iron (Fe)	35–49	50–300	>300
Manganese (Mn)	25–49	50–300	>300
Molybdenum (Mo)	0.40–0.7	0.80–3.3	>3.5
Zinc (Zn)	18–24	25–200	>200

Source: Mills and Jones, 1996.

Table 56   Elemental concentrations in deficient, normal, and toxic cucumber leaves

	% Dry Weight		
Element	Deficient (less than)	Normal Range	Toxic (more than)
Major Elements			
Nitrogen (N), total	—	2.5–5.0	—
Nitrate-N (NO_3-N)	0.4	0.8–1.8	—
Phosphorus (P)	0.3	0.5–1.0	—
Potassium (K)	1.5	3.0–6.0	—
Calcium (Ca)	2.0	2.0–8.0	—
Magnesium (Mg)	0.3	0.4–0.8	—
Sulfur (S)	0.4	0.4–0.8	—

Table 56   Elemental concentrations in deficient, normal, and toxic cucumber leaves (continued)

Element	ppm, Dry Weight		
	Deficient (less than)	Normal Range	Toxic (more than)
Micronutrients			
Boron (B)	30	40–100	200
Iron (Fe)	*	90–150	—
Manganese (Mn)	40	50–300	550
Copper (Cu)	2	4–10	—
Zinc (Zn)	20	50–150	650
Molybdenum (Mo)	0.3	1–3	—

* Not diagnostic.

Source: Gerber, 1985.

Table 57   Elemental concentrations in whole head of healthy greenhouse lettuce and from plants with deficiency and toxicity symptoms

Element	Concentration in Whole Head (%)		
	Deficiency	Sufficiency	Toxicity
Major Elements			
Nitrogen (N)	–	2.1–5.6	–
Phosphorus (P)	<0.6	0.4–0.9	–
Potassium (K)	<3.9	3.9–9.8	–
Calcium (Ca)	<0.8	0.9–2.0	–
Magnesium (Mg)	<0.3	0.4–0.9	–
		(ppm)	
Micronutrients			
Boron (B)	<22	22–65	>54
Copper (Cu)	<2.5	5–17	–
Iron (Fe)	–	56–560	–
Manganese (Mn)	<22	30–200	>200
Zinc (Zn)	<26	33–196	>392

Source: Roorda van Eysinga and Smith, 1981.

Table 58 Elemental concentrations in wrapper leaves of mature Boston lettuce for diagnostic evaluation

Element	Concentration in Wrapper Leaves (%)		
	Low	Sufficient	High
Major Elements			
Nitrogen (N)	2.5–3.9	4.0–5.0	>5.0
Phosphorus (P)	0.2–0.3	0.4–0.6	>0.6
Potassium (K)	5.5–5.9	6.0–7.0	>7.0
Calcium (Ca)	1.2–2.2	2.3–3.5	>3.5
Magnesium (Mg)	0.3–0.4	0.5–0.8	>0.8
		(ppm)	
Micronutrients			
Boron (B)	20–24	25–60	>60
Copper (Cu)	5–7	8–25	>25
Iron (Fe)	40–49	50–100	>100
Manganese (Mn)	10–14	15–250	>250
Zinc (Zn)	20–24	25–250	>250

Source: Mills and Jones, 1996.

Table 59 Elemental concentrations in deficient, normal, and toxic lettuce leaves

Element	% Dry Weight		
	Deficient (less than)	Normal Range	Toxic (more than)
Major Elements			
Nitrogen (N), total	—	2.1–5.6	—
Nitrate-N (NO_3-N)	—	2.5–9.3	—
Phosphorus (P)	0.4	0.5–0.9	—
Potassium (K)	4.0	4.0–10.0	—
Calcium (Ca)	0.8	0.9–2.0	—
Magnesium (Mg)	0.3	0.4–0.8	—
Sulfur (S)	0.2	0.2–0.5	—

Table 59 Elemental concentrations in deficient, normal, and toxic lettuce leaves (continued)

Element	*ppm, Dry Weight*		
	Deficient (less than)	*Normal Range*	*Toxic (more than)*
Micronutrients			
Boron (B)	22	25–65	300
Iron (Fe)	*	50–500	—
Manganese (Mn)	22	25–200	250
Copper (Cu)	2.5	5–18	—
Zinc (Zn)	25	30–200	350
Molybdenum (Mo)	0.2	0.5–3	—

* Not diagnostic.

Source: Gerber, 1985.

Table 60 Elemental concentrations in leaf of tomato plant in mid-bloom—first cluster for diagnostic evaluation

Element	*Concentration in Leaf Opposite and Below Top Flower Cluster (mid-bloom—first cluster) (%)*		
	Low	*Sufficient*	*High*
Major Elements			
Nitrogen (N)	3.00–3.4	3.50–5.0	>5.0
Phosphorus (P)	0.50–0.6	0.70–0.8	>0.8
Potassium (K)	2.00–2.9	3.00–6.0	>6.0
Calcium (Ca)	1.00–1.3	1.40–3.0	>3.0
Magnesium (Mg)	0.25–0.3	0.31–0.7	>0.7
		(ppm)	
Micronutrients			
Boron (B)	23–24	25–75	>75
Copper (Cu)	3–4	5–50	>50
Iron (Fe)	50–59	60–300	>300
Manganese (Mn)	40–49	50–250	>250
Zinc (Zn)	18–19	20–200	>200

Source: Mills and Jones, 1996.

Table 61   Elemental concentrations in leaf of tomato plant in mid-bloom—second cluster for diagnostic evaluation

	Concentration in Leaf Opposite and Below Top Flower Cluster (mid-bloom—second cluster) (%)		
Element	*Low*	*Sufficient*	*High*
Major Elements			
Nitrogen (N)	2.80–3.2	3.21–4.5	>4.5
Phosphorus (P)	0.40–0.5	0.50–0.8	>0.8
Potassium (K)	4.00–5.0	5.01–7.0	>7.0
Calcium (Ca)	1.10–2.0	2.21–4.0	>4.0
Magnesium (Mg)	0.26–0.3	0.31–0.8	>0.8
		(ppm)	
Micronutrients			
Boron (B)	23–24	25–75	>75
Copper (Cu)	3–4	5–50	>50
Iron (Fe)	50–59	60–300	>300
Manganese (Mn)	40–49	50–250	>250
Zinc (Zn)	18–19	20–200	>200

Source: Mills and Jones, 1996.

Table 62   Elemental concentrations in leaf of tomato plant in mid-bloom—third cluster for diagnostic evaluation

	Concentration in Leaf Opposite and Below Top Flower Cluster (mid-bloom—third cluster) (%)		
Element	*Low*	*Sufficient*	*High*
Major Elements			
Nitrogen (N)	2.80–3.0	3.01–4.0	>4.0
Phosphorus (P)	0.40–0.5	0.50–0.8	>0.8
Potassium (K)	4.00–5.0	5.01–7.0	>7.0
Calcium (Ca)	1.10–2.0	2.21–4.0	>4.0
Magnesium (Mg)	0.26–0.3	0.31–0.8	>0.8

Table 62   Elemental concentrations in leaf of tomato plant in mid-bloom—third cluster for diagnostic evaluation (continued)

| Element | *Concentration in Leaf Opposite and Below Top Flower Cluster (mid-bloom—third cluster) (ppm)* | | |
	Low	*Sufficient*	*High*
Micronutrients			
Boron (B)	23–24	25–75	>75
Copper (Cu)	3–4	5–50	>50
Iron (Fe)	50–59	60–300	>300
Manganese (Mn)	40–49	50–250	>250
Zinc (Zn)	18–19	20–200	>200

Source: Mills and Jones, 1996.

Table 63   Elemental concentrations in leaf of tomato plant in mid-bloom—fourth cluster for diagnostic evaluation

| Element | *Concentration in Leaf Opposite and Below Top Flower Cluster (mid-bloom—fourth cluster) (%)* | | |
	Low	*Sufficient*	*High*
Major Elements			
Nitrogen (N)	2.00–2.3	2.31–3.5	>3.5
Phosphorus (P)	0.40–0.5	0.50–0.8	>0.8
Potassium (K)	4.00–5.0	5.01–7.0	>7.0
Calcium (Ca)	1.10–2.0	2.21–4.0	>4.0
Magnesium (Mg)	0.26–0.3	0.31–0.8	>0.8
		(ppm)	
Micronutrients			
Boron (B)	23–24	25–75	>75
Copper (Cu)	3–4	5–50	>50
Iron (Fe)	50–59	60–300	>300
Manganese (Mn)	40–49	50–250	>250
Zinc (Zn)	18–19	20–200	>200

Source: Mills and Jones, 1996.

Table 64 Elemental concentrations in leaf of tomato plant in mid-bloom—fifth cluster for diagnostic evaluation

	Concentration in Leaf Opposite and Below Top Flower Cluster (mid-bloom—fifth cluster) (%)		
Element	Low	Sufficient	High
Major Elements			
Nitrogen (N)	1.70–2.0	2.01–3.0	>3.0
Phosphorus (P)	0.40–0.5	0.50–0.8	>0.8
Potassium (K)	3.00–4.0	4.01–6.0	>6.0
Calcium (Ca)	1.10–2.0	2.21–4.0	>4.0
Magnesium (Mg)	0.26–0.3	0.31–0.8	>0.8
		(ppm)	
Micronutrients			
Boron (B)	23–24	25–75	>75
Copper (Cu)	3–4	5–50	>50
Iron (Fe)	50–59	60–300	>300
Manganese (Mn)	40–49	50–250	>250
Zinc (Zn)	18–19	20–200	>200

Source: Mills and Jones, 1996.

Table 65 Elemental concentrations in leaf of tomato plant in mid-bloom—sixth cluster for diagnostic evaluation

	Concentration in Leaf Opposite and Below Top Flower Cluster (mid-bloom—sixth cluster) (%)		
Element	Low	Sufficient	High
Major Elements			
Nitrogen (N)	1.70–2.0	2.01–3.0	>3.0
Phosphorus (P)	0.40–0.5	0.50–0.8	>0.8
Potassium (K)	3.00–4.0	4.01–6.0	>6.0
Calcium (Ca)	1.10–2.0	2.21–4.0	>4.0
Magnesium (Mg)	0.26–0.3	0.31–0.8	>0.8

Table 65   Elemental concentrations in leaf of tomato plant in mid-bloom—sixth cluster for diagnostic evaluation (continued)

Element	Concentration in Leaf Opposite and Below Top Flower Cluster (mid-bloom—sixth cluster) (ppm)		
	Low	*Sufficient*	*High*
Micronutrients			
Boron (B)	23–24	25–75	>75
Copper (Cu)	3–4	5–50	>50
Iron (Fe)	50–59	60–300	>300
Manganese (Mn)	40–49	50–250	>250
Zinc (Zn)	18–19	20–200	>200

Source: Mills and Jones, 1996.

Table 66   Elemental concentrations in deficient, normal, and toxic tomato leaves

Element	% Dry Weight		
	Deficient (less than)	*Normal Range*	*Toxic (more than)*
Major Elements			
Nitrogen (N), total	2.2	3.0–5.0	—
Nitrate-N (NO_3-N)	4.3	1.2–1.5	—
Phosphorus (P)	0.2	0.4–0.8	—
Potassium (K)	2.0	4.0–8.0	—
Calcium (Ca)	0.7	1.5–4.0	—
Magnesium (Mg)	0.2	0.4–1.0	—
Sulfur (S)	0.5	1.0–3.0	—
	ppm Dry Weight		
Micronutrients			
Boron (B)	30	30–100	150
Iron (Fe)	*	50–150	—
Manganese (Mn)	25	50–200	500
Copper (Cu)	4	5–25	—
Zinc (Zn)	15	20–100	300
Molybdenum (Mo)	0.2	1–5	—

* Not diagnostic.

Source: Gerber, 1985.

Table 67   Criteria for interpretation of foliar analysis of greenhouse tomato

| | Normal Range (% dry weight) | |
Element	Before Fruiting	Fruiting
Major Elements		
Nitrogen (N)	4.5–5.0	3.5–4.0
Phosphorus (P)	0.5–0.8	0.4–0.6
Potassium (K)	3.5–5.0	3.0–4.0
Calcium (Ca)	0.9–2.0	1.0–2.0
Magnesium (Mg)	0.5–1.0	0.4–1.0
Sulfur (S)	0.3–0.8	0.3–0.8
	(ppm, dry weight)	
Micronutrients		
Boron (B)	33–60	35–60
Copper (Cu)	8–20	8–20
Iron (Fe)	50–200	50–200
Manganese (Mn)	50–125	50–125
Zinc (Zn)	25–100	25–100

Source: Faulkner, 1993.

ranges relate to a specific plant part taken at a designated time period or stage of growth. Therefore, these interpretative ranges are not applicable to other types of plant tissues taken at a time other than that given in these tables. This is why it is important to follow given sampling instructions, so that the analytical results obtained can be interpreted based on established sufficiency ranges.

Tissue Testing

Plant analysis is basically defined as a testing method for determining the total elemental content based on a laboratory analysis of collected plant tissue, whereas a tissue test is conducted on extracted plant sap or an extraction of a particular plant part; the test is done on-site using specially designed kits, such as the PLANT CHECK kit shown in Figure 21. For

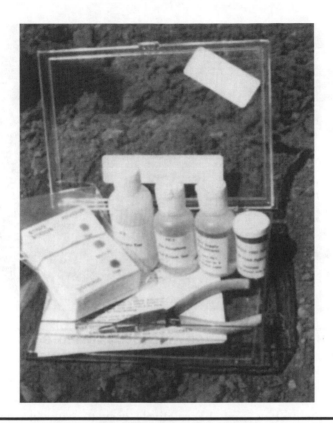

Figure 21   The PLANT CHECK Tissue Testing kit (Spectrum Technologies, Inc., 12010 South Aero Drive, Plainfield, IL 60544).

more comprehensive tests, a HACH kit (shown in Figure 22) can be used. The test is usually made using conductive tissue, such a petioles, leaf midribs, or plant stalks. The commonly determined elements are:

- Nitrogen as nitrate (NO_3^-)
- Phosphorus as phosphate (PO_4^{3-})
- Potassium (K^+)
- Iron (Fe^{3+})

These "quick tests," as they are frequently referred to, can be useful in certain circumstances (Scaife and Stevens, 1983), but they are not to be used as substitutes for a laboratory-conducted analysis. Although the test

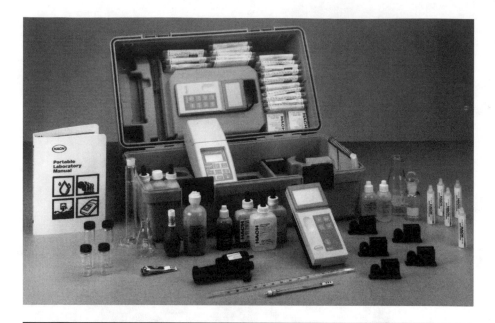

Figure 22   HACH Plant Testing kit (HACH Company, P.O. Box 389, Loveland, CO 80539).

procedures themselves may be relatively easy to conduct, the difficulty comes in interpreting the results, as it takes considerable skill and practice to be able to use tissue test results effectively. In addition, the user needs to have suitable standards available to ensure that the results obtained are analytically correct.

Analytical Devices

With the rapid developments that are being made in all aspects of analytical and diagnostic chemistry, various types of testing kits and devices are coming into the marketplace which can have application for the hydroponic/soilless culture grower. For example, the single-body pH and conductivity meters are now readily available and at a very reasonable cost (Figure 23).

Hand-held specific ion meters, such as the Cardy Nitrate Meter (Figure 24), are very useful for accurately determining the nitrate-nitrogen (NO_3-N)

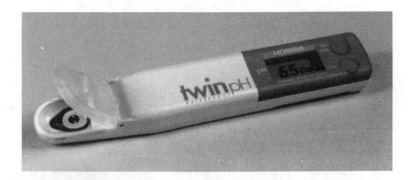

Figure 23   Single-body pH (upper) and conductivity meters (lower). (Courtesy of Spectrum Technologies, Inc., 23839 West Andrew Road, Plainfield, IL 60544.)

content in water, nutrient solutions, or plant tissue sap or extract. Similar meters for the determination of other ions, such as potassium (K^+) and calcium (Ca^{2+}), are being developed, and some are already on the market. However, many of these meters have significant limitations in terms of the types of solution that can be assayed. As stated earlier, reliable standards and solutions of known composition need to be on hand to verify an analysis made by these devices.

The Minolta SPAD 502 Chlorophyll Meter (Figure 25) is finding wide use for estimating the nitrogen content of leaf tissue (Wood et al., 1993), although is application for commonly grown greenhouse crops has not been explored.

Detailed descriptions of various tissue testing techniques have been provided by Jones (1993b) as well as shown in video form (Jones, 1993c).

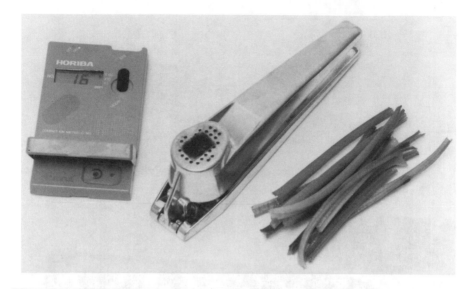

Figure 24   The Cardy Nitrate Meter, garlic press, and plant petioles. (Courtesy of Spectrum Technologies, Inc., 23839 West Andrew Road, Plainfield, IL 60544.)

Summary

It is common practice to focus on single-element deficiencies when dealing with nutritional problems in plants. Since intensive plant production is the nature of most hydroponic/soilless growing systems, equal attention should be given to excesses and imbalances in the concentration of any given element (Berry and Wallace, 1981). This is particularly important with hydroponic systems, where nutrient solution management is so critical to success. Careful monitoring of the nutrient solution as well as the plants themselves should be the normal practice. It is much easier to catch a potential nutrient element problem in its initial stages than to correct it when symptoms appear. The hydroponic/soilless culture grower always needs to be prepared to meet any difficulty with the tools required to solve a problem. Laboratory mailing kits and the required containers should be on hand. If tests are to be conducted in-house, then the kits and testing devices required should be in good working order, with fresh reagents and standards on hand.

Figure 25   The Minolta SPAD 502 Chlorophyll Meter. (Courtesy of Spectrum Technologies, 23839 West Andrew Road, Plainfield, IL 60544.)

With the increasing complexity and the many facets of most growing systems, proper management may be beyond the ability of any one person. Therefore, the hydroponic/soilless culture grower needs to know to whom to turn when important decisions are to be made and/or when a problem arises. Assistance may be provided by a well-trained and experienced county agent, crop consultant, or supplier, but it is important that prior contact be made with such individuals so that time is not lost when a timely decision needs to be made.

Pest Control   12

There is nothing unique about hydroponic/soilless growing with respect to pest control. The same procedures used for growing in soil must be practiced to avoid disease and insect problems. In fact, control measures are more important for the hydroponic grower, as the entire system provides an ideal environment for various types of pests unless measures are taken to control them.

In most instances, good pest control is based on common sense—keeping growing and working areas clean and using good sanitation practices. Keeping pests out of a crop is easier than attempting to control them after they have made their appearance. The trend today is away from chemical control, if at all possible. Since pest control is highly specific, depending on the crop and method of growing, only general recommendations can be given in this text. However, general recommendations can go a long way in preventing the occurrence of a pest problem, thereby avoiding the hazards of a lost crop or the expense of continuous chemical measures in order to maintain some level of control.

The kinds of pest problems a grower may confront and their control vary considerably from one geographic area to another. It is essential that the grower become familiar with the commonly occurring pests in the area and with the recommended control measures. For chemical control, the recommended pesticides and fungicides should be on hand and application equipment made ready for use at the first sign of a problem which may have economic consequences if not quickly brought under control. Daily monitoring procedures must be developed and routinely practiced. It is important to be familiar with those levels of pest incidence considered damaging and therefore economically important to control. Every grower must be able to recognize at what level a pest can be tolerated and therefore requires no treatment.

Integrated Pest Management

For most pest control situations, the grower is advised to develop an integrated pest management (IPM) program, which is a system of monitoring both plant and pest populations using a variety of controls—chemical, cultural, and/or biological—in order to keep plant damage below economic levels (McEno, 1994; Ferguson, 1996; Waterman, 1996).

Sanitation

Sanitation is by far the simplest and most important pest control procedure one can adopt. Since most pest problems are "brought to" the crop, preventing their entrance lies at the root of a good pest management program. Prevention includes using "clean" or sterilized containers, plants, water, growth media, etc. It means keeping the growing area free of foreign plants. Tools, equipment, materials (including clothing), hands, and footwear must be kept free of disease organisms. The vast majority of pest problems are preventable if such procedures become routine practice.

Maintenance of the area around the growing area or greenhouse is equally important. There may be plants in the immediate vicinity of the greenhouse that provide breeding ground for insects and diseases which are then carried into the growing area or greenhouse by wind or human activity. It may be necessary to examine the surrounding area as much as a mile or more, looking upwind first. The installation of wind breaks or wind diversions may be the simplest way to solve a downwind problem.

Prevention Procedures

Chemical-based prevention procedures are also important when dealing with pest problems known to be of common occurrence. For example, the best practice may be to keep plants "covered" with a fungicide to prevent commonly occurring fungus diseases from gaining a foothold. Maintaining specific spray or fumigation schedules may also be good practices in order to keep insect populations under control. Waiting until there are signs of disease or insect pressures may be too late to regain the upper hand. An equally common practice is to vary the type of chemicals applied to prevent the development of pest immunity.

Today, the trend is toward "natural" control of insect pests using predators. Again, success is based on having the predator present in the greenhouse rather than waiting until an insect is seen in large numbers. Passive measures, such as "yellow sticky boards" and insect traps, can be used to provide some degree of control as well as to alert the grower to what is present and at what population level. The application of a pest chemical or other measures may not be economically sound, based on insect counts and the costs to control versus possible crop losses.

Because some diseases are carried by insects, plant infection can be prevented by controlling the insect vector. Therefore, it is important to know the disease cycle and how it is carried from one plant to another, because effective control can be obtained by interrupting any one of the steps in the cycle.

Cultivar Selection

Another very important means of pest control is the selection and use of resistant cultivars (Mohyuddin, 1985). Many of the more common plant diseases that once plagued growers have been essentially eliminated by new cultivars that have been bred specifically for disease resistance. New cultivars are introduced almost every year, and the grower must be aware of them. They offer one of the best means of disease control. Bacterial and viral diseases are best controlled by selecting resistant cultivars.

Environmental Conditions and Cultural Practices

Control of pest problems that occur as a secondary effect becomes difficult or ineffective until the primary cause is identified and corrected. Such is the case, for example, in induced pest problems that gain a foothold because the crop is under elemental or environmental stress. Elemental deficiencies and water and temperature stresses can set up a crop for invasion by some ever-present, but not usually seen, pest. Older plant tissues become easy targets for some types of plant diseases and a desirable habitat for insects. A grower frequently finds a pest management program ineffective until the growing system is sufficiently well managed to control elemental and environmental stresses.

Therefore, it becomes important to determine the primary cause of a developing pest problem so that the correct action can be taken to regain control. Some environmental and plant species associations make hydroponic growing difficult. For example, in warm climates, roots of tomato in gravel or solution media are easily attacked by the fungus *Pythium aphanidermatum*. Most chemical and other techniques are normally ineffective for adequate control of this fungus disease, which forces the grower to select another crop. It is not unusual for the grower to have an excellent first crop free of disease infestation, only to find succeeding crops of the same species increasingly attacked by disease. Complete sterilization between crops of the entire system with steam or a chemical sterilant can eliminate disease, but the cost is high.

Common disease pests are the various fungi that inhabit plant leaves. They vary in type and occurrence, depending on the plant species and environmental conditions. These diseases are particularly severe when environmental conditions are warm and moist. Therefore, the essential control measures are devoted to keeping plant foliage dry and avoiding extremes in temperature. In some instances, keeping plant leaf surfaces covered with a recommended fungicide is required for control in order to keep the disease from gaining a foothold. As mentioned earlier, prevention is far more economical and effective than attempting to bring an infestation under control.

Another aspect of growing which affects the extent of pest infestations relates to the density of the plant canopy. Dense plant canopies make an ideal habitat for many insects and diseases. The penetration of pest chemicals is commonly inhibited by the foliage, leaving areas of leaf surfaces uncovered, with temperature and humidity often ideal for the regeneration and rapid growth of pests. By keeping the plant canopy open, accomplished by proper plant spacing, staking and pruning, providing air movement, and reducing humidity, a less than ideal environment for these pests is created.

The Nutrient Solution

The nutrient solution is an ideal environment for the growth of algae and other pests. Minimizing exposure of the nutrient solution to light can prevent the growth of algae, which if given a foothold in the nutrient solution will clog delivery tubes, pipes, and valves. Filters of various kinds can remove suspended substances from the nutrient solution. Millipore filtering (Millipore Corporation, Ashby Road, Bedford, MA 01730) will, to some extent, remove some disease-producing organisms from the solution. Some pest control chemicals can be added to the nutrient solution to control disease; however, great care is required to keep the concentration at levels that will provide pest control but not harm the crop.

Chemical Use Regulations

In the United States and many other countries, the sale and use of pest chemicals are carefully regulated. Licenses are required to purchase and/or use most pest control chemicals. The label plays an essential role in providing both information on crops that are cleared for use and application procedures. Those who violate these regulations and label clearances are subject to stiff penalties. The greatest health concern arises in connection with food crops, where residues left on the edible portion may be hazardous. Because the laws and label clearances are constantly changing, the grower needs to be sure that use of a particular pest chemical is legal. The best source for current information on pest chemical use is the Agricultural Cooperative Extension Service in the United States and similar governmental agencies elsewhere.

How to Control Pests

It is evident that an effective pest control program includes many elements: (1) use of good sanitation and cultural practices, (2) selection of resistant cultivars, and (3) chemical control. The best pest management program is based on prevention rather than control after infestation.

A pest problem must be properly identified before any corrective step is taken. In certain cases, it may be necessary to call on a trained expert to assist in the identification and to prescribe pest control measures. Sources for such assistance should be identified and located for quick reference when needed.

Insect control can be accomplished by several means. Insects are normally brought to the crop, and it is usually their progeny that do the damage. Knowing something of the life cycle of the insect pest can pinpoint the step easiest to interrupt. Therefore, as with most diseases, prevention is more effective than attempting to bring a damaging population under control. Knowing that an insect infestation might occur and the conditions suitable for it, the grower can then use the appropriate means for control even though the present population is insufficient to damage the crop. For example, white fly is a very common pesky insect which, if given a chance to gain a foothold, will damage a crop quickly, before control measures become effective. With this pest, active (chemical and predator) and passive (yellow sticky boards) control measures must be in place at all times.

One aspect of insect control that varies to some degree compared to disease control is that some level of presence can be tolerated without the need for chemical action. This aspect does require some knowledge of insects and how to judge when control is or is not needed. In some instances, expert advice is required to make this decision. Since most growers are generally not sufficiently expert in pest identification and control, the use of consultants to assist in developing an effective pest management program is important. It is knowing what to expect, what to do, and how to do it that can keep pests out or salvage a crop if a pest gets in.

References

Adi Limited. 1982. Aeroponics in Israel. *HortSci.* 17(2):137.

Ames, M. and W.S. Johnson. 1986. A review of factors affecting plant growth, Section V. In: *Proceedings 7th Annual Conference on Hydroponics: The Evolving Art, The Evolving Science.* Hydroponic Society of America, Concord, CA.

Anon. 1984. *Bibliography on Soilless Culture, 1984.* International Society for Soilless Culture, Wageningen, The Netherlands.

Anon. 1992. *Soil and Plant Analysis Laboratory Registry for the United States and Canada.* Soil and Plant Analysis Council, St. Lucie Press, Delray Beach, FL.

Anon. 1997. Water: What's in it and how to get it out. *Today's Chem.* 6(1):16–19.

Antkowiak, R.J. 1993. More oxygen for your NFT. *The Growing Edge* 4(3):59–62.

Arnon, D.I. and P.R. Stout. 1939. The essentiality of certain elements in minute quantity for plants with special reference to copper. *Plant Physiol.* 14:371–375.

Asher, C.J. 1991. Beneficial elements, functional nutrients, and possible new essential elements, pp. 703–723. In: J.J. Mortvedt (Ed.), *Micronutrients in Agriculture.* SSSA Book Series Number 4. Soil Science Society of America, Madison, WI.

Asher, C.J. and D.G. Edwards. 1978a. Critical external concentrations for nutrient deficiency and excess, pp. 13–28. In: A.R. Ferguson, B.L. Bialaski, and J.B. Ferguson (Eds.), *Proceedings 8th International Colloquium Plant Analysis and Fertilizer Problems.* Information Series No. 134. New Zealand Department of Scientific and Industrial Research, Wellington, New Zealand.

Asher, C.J. and D.G. Edwards. 1978b. Relevance of dilute solution culture studies to problems of low fertility tropical soils, pp. 131–152. In: C.S. Andrew and E.J. Kamprath (Eds.), *Mineral Nutrition of Legumes in Tropical and Subtropical Soils.* Commonwealth Scientific & Industrial Research Organization, Melbourne, Australia.

171

Ashworth, W. 1991. *The Encyclopedia of Environmental Studies.* Facts on File, New York.

Baisden, G. 1994. Frankenfood: Bioengineered bonanza of the future, or your worst nightmare come true. *The Growing Edge* 5(4):34–37, 40–42.

Baker, K.F. (Ed.). 1957. *The U.C. System for Producing Healthy Container-Grown Plants.* California Agricultural Experiment Station Manual 23. Berkeley, CA.

Ball, V. (Ed.). 1985. *Ball Red Book: Greenhouse Growing,* 14th edition. Reston Publishing, Reston, VA.

Bar-Akiva, A. 1984. Substitutes for benzidine as H-donors in the peroxidase assay for rapid diagnosis of iron in plants. *Commun. Soil Sci. Plant Anal.* 15:929–934.

Bar-Akiva, A., D.N. Maynard, and J.E. English. 1978. A rapid tissue test for diagnosis of iron deficiencies in vegetable crops. *Hort. Sci.* 13:284–285.

Barber, S.A. 1995. *Soil Nutrient Bioavailability: A Mechanistic Approach,* 2nd edition. John Wiley & Sons, New York.

Barber, S.A. and D.R. Bouldin (Eds.). 1984. *Roots, Nutrient and Water Influx, and Plant Growth.* ASA Special Publication 136. American Society of Agronomy, Madison, WI.

Bauerle, W. 1990. A window into the future in precision nutrient control, pp. 25–27. In: S. Korney (Ed.), *Proceedings of the 11th Annual Conference on Hydroponics.* Hydroponic Society of America, San Ramon, CA.

Belanger, R.R., P.A. Bowen, D.L. Ehret, and J.G. Menzies. 1995. Soluble silicon: Its role in crop and disease management of greenhouse crops. *Plant Dis.* 79(4):329–335.

Bauerle, W., T.H. Short, E. Mora, S. Hoffman, and T. Nantais. 1988. Computerized individual nutrient fertilizer injector. The System. *HortSci.* 23(5):910.

Berry, W.L. 1985. Nutrient solutions and hydroponics. In: *Proceedings of the 6th Annual Conference on Hydroponics.* Hydroponic Society of America, Concord, CA.

Berry, W.L. 1989. Nutrient control and maintenance in solution culture, pp. 1–6. In: S. Korney (Ed.), *Proceedings of the 10th Annual Conference on Hydroponics.* Hydroponic Society of America, Concord, CA.

Berry, M.L. and A. Wallace. 1981. Toxicity: The concept and relationship to the dose response curve. *J. Plant Nutr.* 3:13–19.

Bezdicek, D.F. (Ed.). 1984. *Organic Farming: Current Technology and Its Role in a Sustainable Agriculture.* ASA Special Publication Number 26. American Society of Agronomy, Madison, WI.

Bij, J. 1990. Growing commercial vegetables in rockwool, pp. 18–24. In: S. Korney (Ed.), *Proceedings of the 11th Annual Conference on Hydroponics.* Hydroponic Society of America, Concord, CA.

Bloom, A. 1987. Nutrient requirement changes during plant development, pp. 104–112. In: *Proceedings 8th Annual Conference: Hydroponics—Effective Growing Techniques.* Hydroponic Society of America, Concord, CA.

Boodley, J.W. and R. Sheldrake, Jr. 1972. *Cornell Peat-Lite Mixes for Commercial Plant Growing.* Information Bulletin No. 43. New York College of Agriculture, Cornell University, Ithaca, NY.

Brooke, L.L. 1995a. For the health of your hydroponic crops. *The Growing Edge* 7(1):21–22, 79.

Brooke, L.L. 1995b. A world ahead: The leaders in hydroponic technology. *The Growing Edge* 6(4): 34–39, 70–71.

Brooks, C. 1992. Development of a semi-automated system for production of salad vegetables for use on space station Freedom, pp. 72–76. In: D. Schact (Ed.), *Proceedings of the 13th Annual Conference on Hydroponics.* Hydroponic Society of America, San Ramon, CA.

Brown, P.H., R.M. Welsh, and E.E. Cary. 1987. Nickel: A micronutrient essential for higher plants. *Plant Physiol.* 85:801–803.

Bruce, R.R., J.E. Pallis, Jr., L.A. Harper, and J.B. Jones, Jr. 1980. Water and nutrient element regulation prescription in nonsoil media for greenhouse crop production. *Commun. Soil Sci. Plant Anal.* 11(7):677–698.

Budenheim, D.L. 1991. Plants for water recycling, oxygen regeneration and food production. *Waste Manage. Res.* 9:435–443.

Budenheim, D.L. 1993. Regenerative growing systems, pp. 17–31. In: P. Bates and S. Korney (Eds.), *Proceedings of the 15th Annual Conference on Hydroponics.* Hydroponic Society of America, San Ramon, CA.

Budenheim, D.L., C.L. Straight, M.T. Flynn, and M. Bates. 1995. Controlled environment agriculture at the Amudsen-Scott South Pole Station, Antarctic and CELSS Antarctic Analog Project, pp. 108–124d. In: M. Bates (Ed.), *Proceedings of the 16th Annual Conference on Hydroponics.* Hydroponic Society of America, San Ramon, CA.

Bugbee, B. 1995. Nutrient management in recirculating hydroponics culture, pp. 15–30. In: M. Bates (Ed.), *Proceedings of the 16th Annual Conference on Hydroponics.* Hydroponic Society of America, San Ramon, CA.

Bunt, A.C. 1988. *Media and Mixes for Container-Grown Plants,* 2nd edition. Unwin Hyman, London, England.

Buyanovsky, G., J. Gale, and N. Degani. 1981. Ultra-violet radiation for the inactivation of microorganisms in hydroponics. *Plant and Soil* 60:131–136.

Carlile, W.R. and E. Sweetland. 1983. The use of composted peat-sludge mixtures in horticultural growth media. *Acta Hort.* 150:510–517.

Carson, E.W. (Ed.). 1974. *The Plant Root and Its Environment.* University Press of Virginia, Charlottesville, VA.

Chaney, R.L. 1983. Plant uptake of inorganic waste constituents, pp. 50–76. In: J.F. Parr, P.B. Marsh, and J.M. Kla (Eds.), *Land Treatment of Hazardous Wastes.* Noyes Data Corporation, Park Ridge, NJ.

Clark, R.B. 1982. Nutrient solution: Growth of sorghum and corn in mineral nutrition studies. *J. Plant Nutr.* 5(8):1003–1030.

Collins, W.L. and M.H. Jensen. 1983. *Hydroponics, A 1983 Technology Overview.* Environment Research Laboratory, University of Arizona, Tucson, AZ.

Cooper, A. 1976. *Nutrient Film Technique for Growing Crops.* Grower Books, London, England.

Cooper, A. 1979. *Commercial Applications of NFT.* Grower Books, London, England.

Cooper, A. 1985. New ABC's of NFT, pp. 180–185. In: A.J. Savage (Ed.), *Hydroponics Worldwide: State of the Art in Soilless Crop Production.* International Center for Special Studies, Honolulu, HI.

Cooper, A. 1988. *The ABC of NFT.* Grower Books, London, England.

Cooper, A. 1996. *The ABC of NFT, Nutrient Film Technique.* Casper Publications, Narrabeen, Australia.

DeKorne, J.B. 1992–93. An orchard of lettuce trees: Vertical NFT system. *The Growing Edge* 4(2): 52–55.

Douglas, J.S. 1976. *Advanced Guide to Hydroponics.* Drake Publishers, New York.

Eastwood, T. 1947. *Soilless Growth of Plants,* 2nd edition. Reinhold Publishing, New York.

Edwards, K. 1985. New NFT breakthroughs and future directions, pp. 186–192. In: A.J. Savage (Ed.), *Hydroponics Worldwide: State of the Art in Soilless Crop Production.* International Center for Special Studies, Honolulu, HI.

Edwards, R. 1994. Computer control systems wired to grow. *The Growing Edge* 5(3): 34–38.

Elliott, B. 1989. Commercial trends in hydroponics, pp. 59–66. In: S. Korney (Ed.), *Proceedings of the 10th Annual Conference on Hydroponics.* Hydroponic Society of America, Concord, CA.

Epstein, E. 1972. *Mineral Nutrition of Plants: Principles and Perspectives.* John Wiley & Sons, New York.

Eskew, D.L., R.M. Welsh, and W.A. Norvell. 1984. Nickel in higher plants: Further evidence for an essential role. *Plant Physiol.* 76:691–693.

Evans, R.D. 1995. Control of microorganism in flowing nutrient solutions, pp. 31–43. In: M. Bates (Ed.), *Proceedings of the 16th Annual Conference on Hydroponics.* Hydroponic Society of America, San Ramon, CA.

Eysinga, J.P.N.L.R. van and K.W. Smilde. 1981. *Nutritional Disorders in Glasshouse Tomatoes, Cucumbers, and Lettuce.* Centre for Agricultural Publishing and Documentation, Wageningen, The Netherlands.

Farnhand, D.S., R.F. Hasek, and J.L. Paul. 1985. *Water Quality.* Leaflet 2995. Division of Agriculture Science, University of California, Davis, CA.

Faulkner, S.P. 1993. Leaf analysis: Measuring nutritional status in plants. *The Growing Edge* 5(1):24–28, 67–68.

Ferguson, G. 1996. Management of greenhouse tomato pests: An integrated approach, pp. 26–30. In: *Proceedings of the Greenhouse Tomato Seminar.* ASHS Press, American Society for Horticultural Science, Alexandria, VA.

Fynn, P. and B. Endres. 1994. Hydroponic rose production, pp. 1–10. In: P. Bates and S. Korney (Eds.), *Proceedings of the 15th Annual Conference on Hydroponics.* Hydroponic Society of America, San Ramon, CA.

Geraldson, C.M. 1963. Quantity and balance of nutrients required for best yields and quality of tomatoes. *Proc. Fla. State Hort. Soc.* 76:153–158.

Geraldson, C.M. 1982. Tomato and the associated composition of the hydroponic or soil solution. *J. Plant Nutr.* 5(8):1091–1098.

Gerber, J.M. 1985. Plant growth and nutrient formulas, pp. 58–69. In: A.J. Savage (Ed.), *Hydroponics Worldwide: State of the Art in Soilless Crop Production.* International Center for Special Studies, Honolulu, HI.

Gerber, J.M. 1986. NFT lettuce production, pp. 77–79. In: C.A. Laymon and J.D. Farley (Eds.), *Annual American Greenhouse Vegetable Growers Conference.* September 28–October 1, Westlake, OH. The Ohio State University, Wooster, OH.

Gerhart, K.A. and R.C. Gerhart. 1992. Commercial vegetable production in a perlite system, pp. 35–38. In: D. Schact (Ed.), *Proceedings of the 13th Annual Conference on Hydroponics.* Hydroponic Society of America, San Ramon, CA.

Gericke, W.F. 1929. Aquaculture—A means of crop production. *Am. J. Bot.* 16:862.

Gericke, W.F. 1937. Hydroponics—Crop production in liquid culture media. *Science* 85:177–178.

Gericke, W.F. 1940. *Complete Guide to Soilless Gardening.* Prentice Hall, New York.

Giacomelli, G. 1991. Introduction to the horticultural engineering aspects of hydroponic crop production, pp. 38–49. In: S. Knight (Ed.), *Proceedings of the 12th Annual Conference on Hydroponics.* Hydroponic Society of America, San Ramon, CA.

Gilbert, H. 1979. *Hydroponic and Soilless Cultures, 1969–May 1979.* Quick Bibliography Series—National Agricultural Library (79-02). USDA, Beltsville, MD.

Gilbert, H. 1983. *Hydroponics/Nutrient Film Technique, 1979–1983* (Bibliography). Quick Bibliography Series—National Agricultural Library (83-31). USDA, Beltsville, MD.

Gilbert, H. 1984. *Hydroponics/Nutrient Film Technique.* Quick Bibliography Series—National Agricultural Library (84-56). USDA, Beltsville, MD.

Gilbert, H. 1985. *Hydroponics/Nutrient Film Technique: 1979–85.* Quick Bibliography Series—National Agricultural Library (86-22). USDA, Beltsville, MD.

Gilbert, H. 1987. *Hydroponics/Nutrient Film Technique: 1981–1986.* Quick Bibliography Series—National Agricultural Library (87-36). USDA, Beltsville, MD.

Gilbert, H. 1992. *Hydroponics/Nutrient Film Technique: 1983–1991.* Quick Bibliography Series—National Agricultural Library (92-43). USDA, Beltsville, MD.

Glass, D.M. 1989. *Plant Nutrition: An Introduction to Current Concepts.* Jones and Bartlett Publishers, Boston, MA.

Handreck, H.A. and N.D. Black. 1993. *Growing Media for Ornamental Plants and Turf,* 4th edition. University of New South Wales Press, Sydney, Australia.

Hewitt, E.J. 1966. *Sand and Water Culture Method Used in Study of Plant Nutrition.* Technical Communication No. 22 (revised). Commonwealth Bureau of Horticulture and Plantation Crops, East Malling, Maidstone, Kent, England.

Hoagland, D.R. and D.I. Arnon. 1950. *The Water Culture Method for Growing Plants without Soil.* Circular 347. California Agricultural Experiment Station, University of California, Berkeley, CA.

Hochmuth, G. 1991. Production of greenhouse tomatoes in Florida. In: G. Hochmuth (Ed.), *Florida Greenhouse Vegetables Production Handbook,* Volume 3. University of Florida, Gainesville, FL.

Hochmuth, G. 1996. Greenhouse tomato nutrition and fertilization for southern latitudes, pp. 37–49. In: *Greenhouse Tomato Seminar.* ASHS Press, American Society for Horticultural Science, Alexandria, VA.

Hochmuth, G. and B. Hochmuth. 1996. Challenges for growing tomatoes in warm climates, pp. 34–36. In: *Greenhouse Tomato Seminar.* ASHS Press, American Society for Horticultural Science, Alexandria, VA.

Hurd, R.G. 1985. United Kingdom: Current research and developments in soilless culture with particular reference to NFT, pp. 164–171. In: A.J. Savage (Ed.), *Hydroponics Worldwide: State of the Art in Soilless Crop Production.* International Center for Special Studies, Honolulu, HI.

Hurd, R.G., P. Adams, D.M. Massey, and D. Price (Eds.). 1980. *Symposium on Re-*

search on Recirculating Water Culture. Acta Horticulture No. 98. The Hague, The Netherlands.

Ikeda, H. and T. Osawa. 1981. Nitrate- and ammonium-N absorption by vegetables from nutrient solution containing ammonium nitrate and the resultant change of solution pH. *Japan. Soc. Hort. Sci.* 50(2):225–230.

Ingratta, F.J., T.J. Blom, and W.A. Straver. 1985. Canada: Current research and developments, pp. 95–102. In: A.J. Savage (Ed.), *Hydroponics Worldwide: State of the Art in Soilless Crop Production.* International Center for Special Studies, Honolulu, HI.

Jensen, M.H. 1981. New developments in hydroponic systems: Descriptions, operating characteristics, evaluation, pp. 1–25. In: *Proceedings: Hydroponics: Where Is It Growing?* Hydroponic Society of America, Brentwood, CA.

Jensen, M.H. 1995. Hydroponics of the future, pp. 125–132. In: M. Bates (Ed.), *Proceedings of the 16th Annual Conference on Hydroponics.* Hydroponic Society of America, San Ramon, CA.

Jensen, M.H. and M.A. Tern. 1971. Use of controlled environments for vegetable production in desert regions of the world. *HortSci.* 6:33–34.

Jones, J.B., Jr. 1980. Construct your own automatic growing machine. *Popular Science* 216(3):87.

Jones, J.B., Jr. 1983. *A Guide for the Hydroponic and Soilless Culture Grower.* Timber Press, Portland, OR.

Jones, J.B., Jr. 1993a. Grower application of media and tissue analysis, pp. 7–14. In: T. Alexander (Ed.), *Proceedings of the 14th Annual Conference on Hydroponics.* Hydroponic Society of America, San Ramon, CA.

Jones, J.B., Jr. 1993b. *Plant Nutrition Manual.* Micro-Macro Publishing, Athens, GA.

Jones, J.B., Jr. 1993c. *Plant Nutrition Basics, The Major Elements, Micronutrients, Plant Analysis, and Tissue Testing* (VHS Video Series). St. Lucie Press, Delray Beach, FL.

Jones, J.B., Jr. and V.W. Case. 1990. Sampling, handling, and analyzing plant tissue samples, pp. 389–427. In: R.L. Westerman (Ed.), *Soil Testing and Plant Analysis,* 3rd edition. SSSA Book Series Number 3. Soil Science Society of America, Madison, WI.

Jones, J.B., Jr., B. Wolf, and H.A. Mills. 1991. *Plant Analysis Handbook: A Practical Sampling, Preparation, Analysis, and Interpretation Guide.* Micro-Macro Publishing, Athens, GA.

Jutras, M.W. 1979. *Nutrient Solutions for Plants.* Circular 182. South Carolina Agriculture Experiment Station, Clemson, SC.

Khudheir, G.A. and P. Newton. 1983. Water and nutrient uptake by tomato plants grown with the nutrient film technique in relation to fruit production. *Acta Hort.* 44:133.

Killebrew, F. 1996. Greenhouse tomato disease identification and management, pp. 21–25. In: *Greenhouse Tomato Seminar.* ASHS Press, American Society for Horticultural Science, Alexandria, VA.

Knight, S.L. 1989. Maximizing productivity for CELSS using hydroponics, pp. 27–33. In: S. Korney (Ed.), *Proceedings of the 10th Annual Conference on Hydroponics.* Hydroponic Society of America, Concord, CA.

Lindsay, W.L. 1979. *Chemical Equilibria in Soils.* John Wiley & Sons, New York.

Linquist, R.K. 1985. Insect and mite pest control for crops grown under protected cultivation, pp. 51–57. In: A.J. Savage (Ed.), *Hydroponics Worldwide: State of the Art in Soilless Crop Production.* International Center for Special Studies, Honolulu, HI.

Lorenz, O.A. and D.N. Maynard. 1988. *Knott's Handbook for Vegetable Growers,* 3rd edition. John Wiley & Sons, New York.

Martin-Prével, P., J. Gagnard, and P. Gautier (Eds.). 1987. *Plant Analysis: As a Guide to the Nutrient Requirements of Temperate and Tropical Crops.* Lavosier Publishing, New York.

Mengel, K. and E.A. Kirkby. 1987. *Principles of Plant Nutrition,* 4th edition. International Potash Institute, Worblaufen-Bern, Switzerland.

Mertz, W. 1981. The essential trace elements. *Science* 213:1332–1338.

McEno, J. 1994. Hydroponic IPM, pp. 61–66. In: Don Parker (Ed.), *The Best of the Growing Edge.* New Moon Publishing, Corvallis, OR.

Mills, H.A. and J.B. Jones, Jr. 1996. *Plant Nutrition Handbook II.* Micro-Macro Publishing, Athens, GA.

Mohyuddin, M. 1985. Crop cultivars and disease control, pp. 42–50. In: A.J. Savage (Ed.), *Hydroponics Worldwide: State of the Art in Soilless Crop Production.* International Center for Special Studies, Honolulu, HI.

Molyneux, C.J. 1988. *A Practical Guide to NFT.* T. Snap & Co. Ltd., Preston, Lancashire, England.

Munson, R.D. and W.L. Nelson. 1990. Principles and practices in plant analysis, pp. 359–387. In: R.L. Westerman (Ed.), *Soil Testing and Plant Analysis,* 3rd edition. SSSA Book Series Number 3. Soil Science Society of America, Madison, WI.

Nielsen, K.F. 1974. Roots and root temperatures, pp. 293–333. In: E.W. Carson (Ed.), *The Plant Root and Its Environment.* University Press of Virginia, Charlottesville, VA.

Ogden, R.J., F.A. Pokorny, H.A. Mills, and M.G. Dunavent. 1987. Elemental status of pine bark-based potting media. *Hort. Rev.* 9:103–131.

Pais, I. 1992. Criteria of essentiality, beneficiality, and toxicity of chemical elements. *Acta Alimentaria* 21(2):145–152.

Pais, I. and J.B. Jones, Jr. 1997. *The Handbook of Trace Elements.* St. Lucie Press, Boca Raton, FL.

Pallas, J.E., Jr. and J.B. Jones, Jr. 1978. Platinum uptake by horticultural crops. *Plant and Soil* 50:207–212.

Papadopoulos, A.P. 1991. *Growing Greenhouse Tomatoes in Soil and in Soilless Media.* Agricultural Canada Publication 1865/E. Communications Branch, Agricultural Canada, Ottawa, Canada.

Papadopoulos, A.P. 1994. *Growing Greenhouse Seedless Cucumbers in Soil and in Soilless Media.* Agricultural Canada Publication 1902/E. Communications Branch, Agricultural and Agri-Food Canada, Ottawa, Canada.

Paterson, J.A. and D.A. Hall. 1981. A method for studying the influence of nutrition on tomato plant vigour in hydroponic culture. *Hort. Res.* 21:103–106.

Peterson, J.C. 1981. Modify your pH perspective. *Florists' Review* 169(4386):34–35, 92, 94.

Pokorny, F.A. 1979. Pine bark container media—An overview. *Combined Proc. Int. Plant Propagators Soc.* 29:484–495.

Raper, C.D., Jr. 1987. Measurement and control of ionic composition using automated ion chromatography. Design and maintenance of recirculating hydroponic systems. *HortSci.* 22:1000.

Resh, H.M. 1990. A world of soilless culture, pp. 33–45. In: S. Korney (Ed.), *Proceedings of the 11th Annual Conference on Hydroponics.* Hydroponic Society of America, San Ramon, CA.

Resh, H.M. 1995. *Hydroponic Food Production.* Woodbridge Press Publishing, Santa Barbara, CA.

Reuter, D.J. and J.B. Robinson (Eds.). 1986. *Plant Analysis: An Interpretation Manual.* Inkata Press Pty. Ltd., Victoria, Australia.

Rodriguez de Cianzio, S.R. 1991. Recent advances in breeding for improving iron utilization by plants, pp. 83–88. In: Y. Chen and Y. Hadar (Eds.), *Iron Nutrition and Interactions in Plants.* Kluwer Academic Publishers, Dordrecht, The Netherlands.

Rogan, M. and M. Finnemore. 1992. The last place on earth: Greenhouse gardening at the South Pole. *The Growing Edge* 3(4):36–38.

Roorda van Eysinga, J.P.N.L. and K.W. Smith. 1981. *Nutritional Disorders in Glass- house Tomatoes, Cucumbers, and Lettuce.* Centre for Agricultural Publishing and Documentation, Wageningen, The Netherlands.

Rorabaugh, P.A. 1995. A brief and practical trek through the world of hydroponics, pp. 7–14. In: M. Bates (Ed.), *Proceedings of the 16th Annual Conference on Hydro- ponics.* Hydroponic Society of America, San Ramon, CA.

Russell, E.J. 1950. *Soil Conditions and Plant Growth.* Longmans, Green and Company, London, England.

Ryall, D. 1993. Growing greenhouse vegetables in a recirculating rockwool system, pp. 35–39. In: T. Alexander (Ed.), *Proceedings of the 14th Annual Conference on Hydroponics.* Hydroponic Society of America, San Ramon, CA.

Sadler, P. 1995. The Antarctic hydroponic project, pp. 97–107. In: M. Bates (Ed.), *Proceedings of the 16th Annual Conference on Hydroponics.* Hydroponic Society of America, San Ramon, CA.

Scaife, A. and K.L. Stevens. 1983. Monitoring sap nitrate in vegetable crops: Compari- son of test strips with electrode methods and effects of time of day and leaf position. *Commun. Soil Sci. Plant Anal.* 14:761–771.

Schippers, P.A. 1979. *The Nutrient Flow Technique.* V.C. Mimeo 212. Department of Vegetable Crops, Cornell University, Ithaca, NY.

Schippers, P.A. 1991. Practical aspects to fertilization and irrigation systems, pp. 14–24. In: S. Knight (Ed.), *Proceedings of the 12th Annual Conference on Hydroponics.* Hydroponic Society of America, San Ramon, CA.

Schon, M. 1992. Tailoring nutrient solution to meet the demands of your plants, pp. 1– 7. In: D. Schact (Ed.), *Proceedings of the 13th Annual Conference on Hydroponics.* Hydroponic Society of America, San Ramon, CA.

Schwartzkopf, S. 1990. Design of an experimental hydroponic system for space flight, pp. 46–56. In: S. Korney (Ed.), *Proceedings of the 11th Annual Conference on Hydroponics.* Hydroponic Society of America, San Ramon, CA.

Sheldrake, R., Jr. and J.W. Boodley. 1965. *Commercial Production of Vegetable and Flower Plants.* Cornell Extension Bulletin 1065. Cornell University, Ithaca, NY.

Smith, I.E. 1985. South Africa: Current research and developments, pp. 150–163. In: A.J. Savage (Ed.), *Hydroponics Worldwide: State of the Art in Soilless Crop Production.* International Center for Special Studies, Honolulu, HI.

Soffer, H. 1985. Israel: Current research and developments, pp. 123–130. In: A.J. Savage (Ed.), *Hydroponics Worldwide: State of the Art in Soilless Crop Production.* Inter- national Center for Special Studies, Honolulu, HI.

Soffer, H. 1988. Research on aero-hydroponics, pp. 69–74. In: *Proceedings of the 9th Annual Conference on Hydroponics.* Hydroponic Society of America, Concord, CA.

Sonneveld, C. 1989. Rockwool as a substrate in protected cultivation. *Chronic Hort.* 29(3):33–36.

Steiner, A.A. 1961. A universal method for preparing nutrient solutions of certain desired composition. *Plant and Soil* 15:134–154.

Steiner, A.A. 1980. The selective capacity of plants for ions and its importance for the composition of the nutrient solution, pp. 37–97. In: R.G. Hurd, P. Adams, D.M. Massey, and D. Price (Eds.), *Symposium on Research on Recirculating Water Culture.* Acta Horticulture No. 98. The Hague, The Netherlands.

Steiner, A.A. 1984. The universal nutrient solution, pp. 633–650. In: *Proceedings Sixth International Congress of Soilless Culture.* The Hague, The Netherlands.

Steiner, A.A. 1985. The history of mineral plant nutrition till about 1860 as source of the origin of soilless culture methods. *Soilless Culture* 1(1):7–24.

Straver, W.A. 1996a. Inert growing media for greenhouse tomatoes, pp. 13–15. In: *Greenhouse Tomato Seminar.* ASHS Press, American Society for Horticultural Science, Alexandria, VA.

Straver, W.A. 1996b. Nutrition of greenhouse tomatoes on inert substrates in northern latitudes, pp. 31–33. In: *Greenhouse Tomato Seminar.* ASHS Press, American Society for Horticultural Science, Alexandria, VA.

Takahashi, E., J.F. Ma, and Y. Miyake. 1990. The possibility of silicon as an essential element for higher plants, pp. 99–122. In: *Comments on Agricultural and Food Chemistry.* Gordon and Breach Scientific Publications, London, England.

Tan, K.H. 1993. *Principles of Soil Chemistry,* 2nd edition. Marcel Dekker, New York.

Tapia, M.L. 1985. Chile and the Antarctic: Current research and developments, pp. 103–155. In: A.J. Savage (Ed.), *Hydroponics Worldwide: State of the Art in Soilless Crop Production.* International Center for Special Studies, Honolulu, HI.

Tibbitts, T.W. 1991. Hydroponic culture of plants in space, pp. 54–60. In: S. Knight (Ed.), *Proceedings of the 12th Annual Conference on Hydroponics.* Hydroponic Society of America, San Ramon, CA.

Trelease, S.F. and H.M. Trelease. 1935. Physiologically balanced culture solutions with stable hydrogen-ion concentration. *Science* 78:438–439.

van Zinderen Bakker, E.M. 1986. Development of hydroponic systems and a look into the future, Section III. In: *Proceedings 7th Annual Conference on Hydroponics: The Evolving Art, the Evolving Science.* Hydroponic Society of America, Concord, CA.

Verwer, F.L. and J.J.C. Wellman. 1980. The possibilities of Grodan rockwool in horticulture, pp. 263–278. In: *Fifth International Congress on Soilless Culture.* International Society for Soilless Culture, Wageningen, The Netherlands.

Wallace, A. 1971. *Regulation of the Micronutrient Status of Plants by Chelating Agents and Other Factors.* UCLA 34P51-33. Arthur Wallace, Los Angeles, CA.

Wallace, A. 1989. Regulation of micronutrients and uses of chelating agents in solution culture, pp. 50–53. In: S. Korney (Ed.), *Proceedings of the 10th Annual Conference on Hydroponics.* Hydroponic Society of America, Concord, CA.

Warnke, D.D. 1986. Analyzing greenhouse growth media by the saturation extraction method. *HortSci.* 21:223–225.

Warnke, D.D. 1988. Recommended test procedure for greenhouse growth media, pp. 34–37. In: W.C. Wahnke (Ed.), *Recommended Chemical Soil Test Procedures.* North Central Regional Publication No. 221 (revised). North Dakota Agricultural Experiment Station, Fargo, ND.

Waterman, M.P. 1993–94. Building a better tomato. *The Growing Edge* 5(2):21–25, 69–70.

Waterman, M.P. 1996. The good, bad, and the ugly. *The Growing Edge* 7(3):22–28, 75.

Waters, W.E., J. Nesmith, C.M. Geraldson, and S.S. Woltz. 1972. *The Interpretation of Soluble Salt Tests and Soil Analysis by Different Procedures.* AREC Mimeo Report GC-1972-4. Bradenton, FL.

White, J.W. 1974. Dillon Research Fund, progress report on research at Penn State. *Pennsylvania Flower Growers Bull.* 89:3–4.

Wilcox, G. 1983. Hydroponic systems around the world, their characteristics and why they are used. In: *Proceedings 4th Annual Conference: Hydroponics—How Does It Work?* Hydroponic Society of America, Concord, CA.

Wilcox, G. 1991. Nutrient control in hydroponic systems, pp. 50–53. In: S. Knight (Ed.), *Proceedings of the 12th Annual Conference on Hydroponics.* Hydroponic Society of America, Concord, CA.

Wood, C.W., D.W. Reeves, and D.G. Himelrick. 1993. Relationships between chlorophyll meter readings and leaf chlorophyll concentration, N status, and crop yield: A review. *Proc. Agron. Soc. N.Z.* 23:1–9.

Appendix A: Glossary

The definitions given here are oriented primarily to the jargon of soilless media culture and hydroponic growing, although some apply broadly to all forms of plant growing and the botanical and horticultural sciences.

Absorptive capacity—A measure of the capacity of a substance used as a growing medium in soilless culture to take (absorb) into pores and cavities nutrient solution. The trapped solution is a potential future source of water and essential elements. The composition of the nutrient solution is unaffected by this absorption. (See also *Adsorptive capacity*)

Acidity—Refers to the pH of the nutrient solution or growth medium when the pH measures less than 7.0. An increasing hydrogen ion concentration leads to increasing acidity as the pH decreases from 7.0. (See also *Alkalinity*)

Active absorption—Refers to the process of ion uptake by plant roots requiring the expenditure of energy. This process is controlled and specific as to the number and type of ion species absorbed. (See also *Passive absorption*)

Adsorptive capacity—A measure of the capacity of a substance used as a growing medium in soilless culture to selectively remove from the nutrient solution essential elements by precipitation, complexing, or ion exchange. Adsorbed elements may be released and therefore available to plants at a later time. The adsorptive capacity of a substance will significantly affect the composition of the nutrient solution through time, depending on the degree of adsorption or release. (See also *Absorptive capacity*)

Aerated standing nutrient solution culture—A method of growing plants hydroponically where the plant roots are suspended in a container of continuously aerated nutrient solution. The usual procedure is to maintain the volume of the solution by daily addition of water and to replace the nutrient solution periodically with a fresh batch.

Aeroponics—A technique for growing plants hydroponically where the plant roots are suspended in a container and are either continuously or periodically bathed in a fine mist of nutrient solution.

Alkalinity—Refers to the pH of the nutrient solution or growth medium when the pH measures greater than 7.0. A decreasing hydrogen ion concentration leads to increasing alkalinity as the pH increases from 7.0. (See also *Acidity*)

Anion—Solution having a negative charge. When applied to the composition of the nutrient solution, it designates ions, such as BO_3^{3-}, Cl^-, $H_2PO_4^-$, HPO_4^{2-}, MoO_4^{2-}, NO_3^-, and SO_4^{2-}, which are common forms for these essential elements in solution. In chemical notation, the minus sign indicates the number of electrons the compound will give up. (See also *Cation*)

Atmospheric demand—The capacity of air surrounding the plant to absorb moisture. This capacity of the air will influence the amount of water transpired by the plant through its exposed surfaces. Atmospheric demand varies with changing atmospheric conditions. It is greatest when air temperature and movement are high and relative humidity is low. The reverse conditions exist when atmospheric demand is low.

Availability—A term used to indicate that an element is in a form and position suitable for plant root absorption.

Bag culture—A technique for growing plants in a bag of soilless media (such as mixtures of sphagnum peat moss, pine bark, vermiculite, perlite, etc.) into which a nutrient solution is applied periodically.

Beneficial elements—Elements not essential for plants but which, when present in the nutrient solution at specific concentrations or in rooting media, enhance plant growth.

Boron (B)—An essential element classed as a micronutrient. Boron exists in the nutrient solution as either the borate (BO_3^{3-}) anion or molecular boric acid (H_3BO_3). The primary chemical source is boric acid.

Buffer capacity—The ability of the nutrient solution or growth medium to resist a change in pH during the period of its use.

Calcium (Ca)—An essential element classed as a major element. Calcium exists in the nutrient solution as the divalent cation (Ca^{2+}). Major chemical sources are calcium nitrate [$Ca(NO_3)_2 \cdot 4H_2O$], calcium chloride ($CaCl_2$), and calcium sulfate ($CaSO_4 \cdot 2H_2O$).

Carbon (C)—An essential element classed as a major element. Carbon is obtained from carbon dioxide (CO_2) in the air fixed during photosynthesis.

Cation—An ion in solution having a positive charge. When applied to the composition of a nutrient solution, it designates ions such as, Ca^{2+}, Cu^{2+}, Fe^{3+}, H^+, K^+, Mg^{2+}, Mn^{2+}, NH_4^+, and Zn^{2+}, which are common forms for these essential elements in solution. In chemical notation, the plus sign indicates the number of electrons the element will accept. (See also *Anion*)

Chelates—A type of chemical compound in which a metallic atom (such as iron) is firmly combined with a molecule by means of multiple chemical bonds. The term refers to the claw of a crab, illustrative of the way in which the atom is held.

Chlorine (Cl)—An essential element classed as a micronutrient. Chlorine exists in the nutrient solution as the monovalent anion (Cl^-). Since the chloride anion is ever-present in the environment, it is not specifically added to the nutrient solution.

Chlorosis—A light-green to yellow coloration of leaves or whole plants which usually indicates an essential element insufficiency or toxicity. Chlorosis is most frequently associated with iron deficiency.

Closed hydroponic system—Designates a circulation system of nutrient solution flow. (See also *Open hydroponic system*)

Continuous flow nutrient solution culture—A method of soilless culture in which the plant roots are continuously bathed in a flowing stream of nutrient solution.

Copper (Cu)—An essential element classed as a micronutrient. Copper exists in the nutrient solution as the cupric cation (Cu^{2+}). The primary chemical source is copper sulfate ($CuSO_4 \cdot 5H_2O$).

Deficiency—Describes the condition when an essential element is not in sufficient supply or proper form to adequately supply the plant or is not in sufficient concentration in the plant to meet the plant's physiological requirement. Plants therefore usually grow poorly and show visual signs of abnormality in color and structure.

Diffusion—The movement of an ion in solution at a high concentration to an area of lower concentration. Movement continues as long as the concentration gradient exists.

Drip nutrient solution culture—A method of soilless culture in which the nutrient solution is slowly applied as drops onto the rooting medium.

Electrical conductivity—A measure of the electrical resistance of a nutrient solution, or effluent from a growing bed or pot, used to determine the level of ions in solution. Conductivity may be expressed as specific conductance in mhos (micro- or milli-) or decisiemens (dS) or as resistance in ohms. (See also *Specific conductance*)

Essential elements—Those elements that are necessary for higher plants to complete their life cycle; also refers to the requirements established for essentiality by Arnon and Stout (see Chapter 5).

Feeding cycle—The time period when the nutrient solution is circulated through the root growing medium in those systems where plant roots are only periodically exposed to the nutrient solution.

Gravel culture—A soilless culture technique where plants are grown in beds containing gravel which are periodically bathed in nutrient solution. The gravel serves as a root support system for the plants.

Hydrogen (H)—An essential element classed as a major element. Hydrogen is obtained from water (H_2O) and is fixed during photosynthesis.

Hydroponics—A word coined in the early 1930s by Dr. W.F. Gericke (a University of California researcher) to describe a soilless technique for growing plants. The word was derived from two Greek words: *hydro,* meaning water, and *ponos,* meaning labor—literally *working water.* Hydroponics has been defined as the science of growing plants without the use of soil, but rather by use of an inert medium to which a nutrient solution containing all the essential elements needed by the plant for normal growth and successful completion of its life cycle is periodically added. In this text, hydroponics refers only to those systems of soilless growing that do not use a rooting medium.

Intermittent flow nutrient solution culture—A method of soilless culture in which the nutrient solution is only periodically brought into contact with plant roots.

Ion—An atom or group of atoms having either a positive or negative charge from having lost or gained one or more electrons. (See also *Anion* and *Cation*)

Ion exchange—A method of water purification in which water is passed through a resin bed to remove both cations and anions from the water. Ion exchange also refers to the phenomenon of physical-chemical attraction between charged colloidal substances with cations and anions. Ions of the essential elements can be removed from or released into the nutrient solution by ion exchange characteristics of sphagnum peat moss, pine bark, vermiculite, and clay colloids adhering to sand and gravel particles.

Iron (Fe)—An essential element classed as a micronutrient. Iron exists in the nutrient solution as either the ferrous (Fe2) or ferric (Fe^{3+}) cation. The major chemical sources are iron tartrate, iron citrate and the chelate form, FeEDTA.

Leaf analysis—A method of determining the total elemental content of a leaf and relating this concentration to the well-being of the plant in terms of its elemental composition. (See also *Plant analysis*)

Magnesium (Mg)—An essential element classed as a major element. Magnesium exists in the nutrient solution as the divalent cation (Mg^{2+}). The primary chemical source is magnesium sulfate ($MgSO_4 \cdot 7H_2O$).

Major essential elements—The nine essential elements found in relatively large concentrations in plant tissues. These elements are calcium, carbon, hydrogen, magnesium, nitrogen, oxygen, phosphorus, potassium, and sulfur.

Manganese (Mn)—An essential element classed as a micronutrient. Manganese exists in the nutrient solution as the manganous cation (Mn^{2+}). The primary chemical source is manganese sulfate ($MnSO_4 \cdot H_2O$).

Mass flow—The movement of ions as a result of the flow of water; the ions are carried in the moving water.

Micronutrients—The seven essential elements required by and found in relatively small concentrations in plant tissue. These elements are boron, chlorine, copper, iron, manganese, molybdenum, and zinc.

Mineral nutrition—The study of the essential elements as they relate to the growth and well-being of plants.

Mist nutrient solution culture—See *Aeroponics*.

Molybdenum (Mo)—An essential element classed as a micronutrient. Molybdenum exists in the nutrient solution as the molybdate (MoO_4^{2-}) anion. The primary chemical source is ammonium molybdate $[(NH_4)_6Mo_7O_{24} \cdot 4H_2O]$.

Necrosis—The dead tissue on plant leaves and stems which results from poor nutrition, disease damage, overheating, etc.

Nitrogen (N)—An essential element classed as a major element. Nitrogen is found in the nutrient solution as either the nitrate (NO_3^-) anion or the ammonium (NH_4^+) cation. The primary chemical sources are ammonium nitrate (NH_4NO_3), potassium or calcium nitrate [KNO_3 and $Ca(NO_3)_2 \cdot 4H_2O$, respectively], ammonium sulfate [$(NH_4)_2SO_4$], and ammonium mono- or dihydrogen phosphate [$(NH_4)_2HPO_4$ and $NH_4H_2PO_4$, respectively]. Urea [$CO(NH_2)_2$] is also a commonly used nitrogen source, but it has only very special uses for the hydroponic and soilless grower.

Nutrient film technique (NFT)—A technique for growing plants hydroponically in which the plant roots are suspended in a slow-moving stream of nutrient solution. The technique was developed by Dr. Allen Cooper.

Nutrient solution—A water solution that contains one or more of the essential elements in suitable form and concentration for absorption by plant roots.

Open hydroponic system—Designates a one-way passage of the nutrient solution through the rooting media or trough. After this single passage, the solution is dumped. (See also *Closed hydroponic system*)

Osmotic pressure—Force exerted by substances dissolved in water which affects water movement into and out of plant cells. The salts dissolved in nutrient solutions exert some degree of force which can restrict water movement into plant root cells or extract water from them.

Oxygen (O_2)—An essential element classed as a major element. Oxygen is obtained from carbon dioxide (CO_2) in the air; it is fixed during photosynthesis.

Passive absorption—The movement of ions into plant roots carried along with water being absorbed by roots. (See also *Active absorption*)

Perlite—An aluminosilicate of volcanic origin. When this natural substance is crushed and heated rapidly to 1000°C, it forms a white, lightweight aggregate with a closed cellular structure. Perlite has an average density of 8 pounds per cubic foot (128 kg/m^3), has virtually no cation exchange capacity, is devoid of plant nutrients, contains some fluoride (17 mg/kg, ppm), and is graded into various particle sizes for use as a rooting medium or added to soilless mixes.

pH—The negative logarithm of the hydrogen ion concentration to the base 10:

$$pH = \log_{10} \times 1/[H^+]$$

As pH is logarithmic, the hydrogen ion concentration in solution increases ten times when the pH is lowered one unit. The pH of the nutrient solution and rooting media will significantly affect the availability and utilization of the essential elements.

Phosphorus (P)—An essential element classed as a major element. Phosphorus exists in the nutrient solution as an anion, either as $H_2PO_4^-$ or HPO_4^{2-}, depending on the pH. The primary chemical sources are ammonium or potassium mono- or dihydrogen phosphate [$(NH_4)_2HPO_4$, K_2HPO_4, $NH_4H_2PO_4$, and KH_2PO_4, respectively] and phosphoric acid (H_3PO_4).

Photosynthesis—The process by which chloroplasts in the presence of light split water (H_2O) and combine with carbon dioxide (CO_2) to form simple carbohydrates and release oxygen (O_2):

$$6CO_2 + 6H_2O \rightarrow \text{chloroplasts in light} \rightarrow C_6H_{12}O_6 + 6O_2$$

Pine bark—A by-product of the processing of pine, usually southern yellow pine, for lumber. Bark stripped from the tree is allowed to age in the natural environment for 6 months to 1 year and is then passed through a 1-

inch screened hammer mill. The resulting material is screened into fractions of various sizes for addition to organic growing mixes. Pine bark has substantial cation exchange and water-holding capacities.

Plant analysis—A method of determining the total elemental content of the whole plant or one of its parts and then relating the concentration found to the well-being of the plant in terms of its elemental requirements. (See also *Leaf analysis*)

Plant nutrients—Those elements that are essential to plants. (See *Major essential elements*; *Micronutrients*)

Plant nutrition—The study of the effects of the essential as well as other elements on the growth and well-being of plants.

Plant requirement—That quantity of an essential element needed for the normal growth and development of the plant without inducing stress from a deficiency or an excess.

Potassium (K)—An essential element classed as a major element. It exists in the nutrient solution as a monovalent (K^+) cation. The primary chemical sources are potassium chloride (KCl) and potassium sulfate (K_2SO_4).

Reverse osmosis—A method of water purification in which ions are removed from water by an electrical potential placed on either side of a membrane which acts to extract ions from a passing stream of water.

Rockwool—An inert fibrous material produced from a mixture of volcanic rock, limestone, and coke, melted at 1500 to 2000°C, extruded as fine fibers, and then pressed into loosely woven sheets. Rock wool has excellent water-holding capacity. For growing uses, the rockwool sheets are formed into slabs or cubes.

Salt index—A relative measure of the osmotic pressure of a solution of a fertilizer material in relation to an equivalent concentration of sodium nitrate ($NaNO_3$) whose salt index is set at 100 (see Table 20).

Sand culture—A soilless culture technique where plants are grown in a bed containing sand, which is periodically bathed in nutrient solution.

Scorch—Burned leaf margins. This visual symptom is typical of potassium deficiency or chloride excess.

Secondary elements—Obsolete term used to classify three of the major essential elements: calcium, magnesium, and sulfur.

Soluble salts—A measure of the concentration of ions in water (or nutrient solution) used to determine the quality of the water or solution, measured in terms of its electrical conductivity. (See also *Specific conductance)*

Specific conductance—The reciprocal of the electrical resistance of a solution, measured using a standard cell and expressed as mhos per centimeter (or decisiemens per meter) at 25°C:

$$\text{Specific Conductance} = \theta/R$$

where θ is the cell constant and R is the resistance in ohms. (See also *Electrical conductivity*)

Sufficiency—The adequate supply of an essential element to the plant; also, an adequate concentration of an essential element in the plant to satisfy the plant's physiological requirement. The plant in such a condition will look normal in appearance, be healthy, and be capable of high production.

Sulfur (S)—An essential element classed as a major element. Sulfur exists in the nutrient solution as the sulfate (SO_4^{2-}) anion. The primary chemical sources are as potassium, magnesium, or ammonium sulfates [K_2SO_4, $MgSO_4 \cdot 7H_2O$, and $(NH_4)_2SO_4$, respectively].

Sump—The reservoir for storage of the nutrient solution in closed, recirculating soilless culture systems.

Tissue testing—A method for determining the concentration of the soluble form of an element in the plant by analyzing sap that has been physically extracted from a particular plant part, usually from stems or petioles. Tests are usually limited to the determination of nitrate, phosphate, potassium, and iron. Tissue tests are normally performed using simple analysis kits, and the elemental concentration found is related to the well-being of the sampled plant.

Toxicity—The condition in which an element is sufficiently in excess in the rooting media, nutrient solution, or plant to be detrimental to the plant's normal growth and development.

Trace element—Once commonly used to designate those essential elements that are currently referred to as micronutrients; designates those elements found in plants at low concentration levels, usually at a few to less than 1 µg/kg (ppm) of the dry weight.

Tracking—A technique of following through time the essential element content of the rooting media or plant by frequent time-spaced analyses.

Valence—The combining capacity of atoms or groups of atoms. For example, potassium (K^+) and ammonium (NH_4^+) are monovalent, whereas calcium (Ca^{2+}) and magnesium (Mg^{2+}) are divalent. Some elements may have more than one valance state, such as iron, which can be either divalent (Fe^{2+}) or trivalent (Fe^{3+}). This change from one valance state to another involves the transfer of an electron.

Vermiculite—An aluminum-iron-magnesium silicate. When heated for about one minute to 1000°C, this plate-like, naturally occurring substance expands to 15 to 20 times its original volume, forming a lightweight, high-porosity material which has a density of about 5 pounds per cubic foot (80 kg/m^3). Vermiculite has a fairly high cation-exchange capacity (100 to 150 meq/100 g) and contains plant-available potassium and magnesium. Normally, vermiculite is added to an organic mix to increase the water-holding capacity of the mix, particularly for germinating mixes.

Zinc (Zn)—An essential element classed as a micronutrient. Zinc exists in the nutrient solution as the divalent cation (Zn^{2+}). The primary chemical source is zinc sulfate ($ZnSO_4 \cdot 7H_2O$).

Appendix B: Characteristics of the Essential Elements

In this appendix, the characteristics of the essential elements are presented in outline form for easy reference. The objective is to provide the most useful information about each essential element in one easy-to-follow format. The information and data given are primarily in reference to the hydroponic/soilless growing methods for those crops thus commonly grown; therefore, the information given may not be useful for application with other growing methods or crops. The critical and excessive levels and the sufficiency ranges for the essential elements have been selected as probable levels and should not be considered specific. These levels are what would be found in recently mature leaves, unless otherwise specified.

Nitrogen (N)

Atomic Number: 7   **Atomic Weight**: 14.00

Discoverer of Essentiality and Year: DeSaussure—1804

Designated Element: Major element

Function: Used by plants to synthesize amino acids and form proteins, nucleic acids, alkaloids, chlorophyll, purine bases, and enzymes

Mobility: Mobile

Forms Utilized: Ammonium (NH_4^+) cation and nitrate (NO_3^-) anion

Common Reagent Sources:

	Formula	% N
Ammonium dihydrogen phosphate	$NH_4H_2PO_4$	11 (21% P)
Ammonium hydroxide	NH_4OH	20–25
Ammonium nitrate	NH_4NO_3	32 (16% NH_4 and 16% NO_3)
Ammonium sulfate	$NH_4(SO_4)_2$	21 (24% S)
Diammonium hydrogen phosphate	$(NH_4)_2HPO_4$	18 (21% P)
Calcium nitrate	$Ca(NO_3)_2 \cdot 4H_2O$	15 (19% Ca)
Potassium nitrate	KNO_3	13 (36% K)

Concentration in Nutrient Solutions: 100–200 mg/L (ppm); 3 to 4 parts nitrate (NO_3) to 2 and 1 parts ammonium (NH_4) for best plant growth

Typical Deficiency Symptoms: Very slow-growing, weak, and stunted plants; leaves light green to yellow in color, beginning with the older leaves; plants will mature early, and dry weight and fruit yield will be reduced

Symptoms of Excess: Plants will be dark green in color with succulent foliage; easily susceptible to environmental stress and disease and insect invasion; poor fruit yield of low quality

Critical Levels: 3.00% total nitrogen (will vary with plant type and stage of growth); 1000 mg/kg (ppm) nitrate (NO_3)-nitrogen in leaf petiole

Sufficiency Range:

Plant	Plant Part	% N
Head lettuce	Whole head	2.1–5.6
Lettuce	Wrapper leaf	4.0–5.0
Bell pepper	Fully developed leaf	3.0–5.0
Tomato	Opposite below flower cluster	
	First cluster	3.5–5.0
	Second cluster	3.2–4.5
	Third cluster	3.0–4.0
	Fourth cluster	2.3–3.5
	Fifth cluster	2.0–3.0
	Sixth cluster	2.0–3.0
Cucumber	Fully developed leaf	4.3–6.0

Excessive Level: >5.00% total nitrogen (will vary with plant type and stage of growth); >12,000 mg/kg (ppm) nitrate (NO_3)-nitrogen in leaf petiole.

Ammonium Toxicity: When ammonium (NH_4) is the major source of nitrogen, toxicity can occur, seen as cupping of plant leaves, breakdown of vascular tissue at the base of the plant, lesions on stems and leaves, and increased occurrence of blossom-end rot on fruit

Phosphorus (P)

Atomic Number: 15   **Atomic Weight**: 30.973

Discoverer of Essentiality and Year: Ville—1860

Designated Element: Major element

Function: Component of certain enzymes and proteins involved in energy transfer reactions and component of RNA and DNA

Mobility: Mobile

Forms Utilized: Mono- and dihydrogen phosphate ($H_2PO_4^-$ and HPO_4^{2-}, respectively) anions, depending on pH

Common Reagent Sources:

	Formula	% P
Ammonium dihydrogen phosphate	$NH_4H_2PO_4$	21 (11% N)
Diammonium hydrogen phosphate	$(NH_4)_2HPO_4$	21 (81% N
Dipotassium hydrogen phosphate	K_2HPO_4	18 (22% K)
Phosphoric acid	H_3PO_4	34
Potassium dihydrogen phosphate	KH_2PO_4	32 (30% K)

Concentration in Nutrient Solutions: 30–50 mg/L (ppm)

Typical Deficiency Symptoms: Slow and reduced growth, with developing purple pigmentation of older leaves; foliage may also appear very dark green in color

Symptoms of Excess: Plant growth will be slow, with some visual symptoms possibly related to a micronutrient deficiency, such as zinc

Critical Level: 0.25% total; 500 mg/kg (ppm) extractable phosphorus in leaf petiole

Sufficiency Ranges:

Plant	Plant Part	% P
Head lettuce	Whole head	0.4–0.9
Lettuce	Wrapper leaf	0.4–0.9
Bell pepper	Fully developed leaf	0.2–0.7
Tomato	Opposite below flower cluster	
	First cluster	0.7–0.8
	Second cluster	0.5–0.8
	Third cluster	0.5–0.8
	Fourth cluster	0.5–0.8
	Fifth cluster	0.5–0.8
	Sixth cluster	0.5–0.8
Cucumber	Fully developed leaf	0.3–0.7

Excessive Level: >1.00% total; >3000 mg/kg (ppm) extractable phosphorus in leaf petiole

Potassium (K)

Atomic Number: 19   **Atomic Weight**: 39.098

Discoverer of Essentiality and Year: von Sachs, Knop—1860

Designated Element: Major element

Function: Maintains the ionic balance and water status in plants; involved in the opening and closing of stomata and is associated with the carbohydrate chemistry

Mobility: Mobile

Form Utilized: Potassium (K^+) cation

Common Reagent Sources:

	Formula	% K
Dipotassium hydrogen phosphate	K_2HPO_4	22 (18% P)
Potassium chloride	KCl	50 (47% Cl)
Potassium dihydrogen phosphate	KH_2PO_4	30 (32% P)
Potassium nitrate	KNO_3	36 (13% N)
Potassium sulfate	K_2SO_4	42 (17% S)

Concentration in Nutrient Solutions: 100–200 mg/L (ppm)

Typical Deficiency Symptoms: Initially slowed growth with marginal death of older leaves giving a "burned" or scorched appearance; fruit yield and quality reduced; fruit post-harvest quality reduced

Symptoms of Excess: Plants will develop either magnesium or calcium deficiency symptoms

Critical Level: 2.00%

Sufficiency Ranges:

Plant	Plant Part	% K
Head lettuce	Whole head	3.9–9.8
Lettuce	Wrapper leaf	6.0–7.0
Bell pepper	Fully developed leaf	3.5–4.5
Tomato	Opposite below flower cluster	
	First cluster	3.0–6.0
	Second cluster	5.0–7.0
	Third cluster	5.0–7.0
	Fourth cluster	5.0–7.0
	Fifth cluster	4.0–6.0
	Sixth cluster	4.0–6.0
Cucumber	Fully developed leaf	2.5–4.0

Excessive Level: >6.00%, which will be less depending on plant type and stage of growth

Calcium (Ca)

Atomic Number: 20 **Atomic Weight**: 40.07

Discoverer of Essentiality and Year: von Sachs, Knop—1860

Designated Element: Major element

Functions: Major constituent of cell walls, for maintaining cell wall integrity and membrane permeability; enhances pollen germination and growth; activates a number of enzymes for cell mitosis, division, and elongation; may detoxify the presence of heavy metals in tissue

Mobility: Immobile

Form Utilized: Calcium (Ca^{2+}) cation

Common Reagent Sources:

	Formula	% Ca
Calcium chloride	$CaCl_2$	36 (64% Cl)
Calcium nitrate	$Ca(NO_3)_2 \cdot 4H_2O$	19 (15% N)
Calcium sulfate	$CaSO_4 \cdot 2H_2O$	23 (19% S)

Concentration in Nutrient Solutions: 200–300 mg/L (ppm)

Typical Deficiency Symptoms: Leaf shape and appearance will change, with the leaf margins and tips turning brown or black; edges of leaves may look torn; vascular breakdown at the base of the plant; for fruit crops, occurrence of blossom-end-rot

Symptoms of Excess: May induce possible magnesium or potassium deficiency

Critical Level: 1.00% (will vary with plant type and stage of growth)

Sufficiency Ranges:

Plant	Plant Part	% Ca
Head lettuce	Whole head	0.9–2.0
Lettuce	Wrapper leaf	2.3–3.5
Bell pepper	Fully developed leaf	1.3–2.8
Tomato	Opposite below flower cluster	
	First cluster	1.4–3.0
	Second cluster	2.2–4.0
	Third cluster	2.2–4.0
	Fourth cluster	2.2–4.0
	Fifth cluster	2.2–4.0
	Sixth cluster	2.2–4.0
Cucumber	Fully developed leaf	2.5–4.0

Excessive Level: >5.00% (will vary with level of potassium and/or magnesium)

Magnesium (Mg)

Atomic Number: 12   **Atomic Weight**: 24.30

Discoverer of Essentiality and Year: von Sachs, Knop—1860

Designated Element: Major element

Functions: Major constituent of the chlorophyll molecule; enzyme activator for a number of energy transfer reactions

Mobility: Moderately immobile

Form Utilized: Magnesium (Mg^{2+}) cation

Common Reagent Sources:

	Formula	% Mg
Magnesium sulfate	$MgSO_4 \cdot 7H_2O$	10 (23% S)

Concentration in Nutrient Solutions: 30–50 mg/L (ppm)

Typical Deficiency Symptoms: Interveinal chlorosis on older leaves; possible development of blossom-end-rot in fruit

Symptoms of Excess: Results in cation imbalance among calcium and potassium; slowed growth with the possible development of either calcium or potassium deficiency symptoms

Critical Level: 0.25%

Sufficiency Ranges:

Plant	Plant Part	% Mg
Head lettuce	Whole head	0.4–0.9
Lettuce	Wrapper leaf	0.5–0.8
Bell pepper	Fully developed leaf	0.3–1.0
Tomato	Opposite below flower cluster	
	First cluster	0.3–0.7
	Second cluster	0.3–0.8
	Third cluster	0.3–0.8
	Fourth cluster	0.3–0.8
	Fifth cluster	0.3–0.8
	Sixth cluster	0.3–0.8
Cucumber	Fully developed leaf	0.5–1.0

Excessive Level: >1.50% (will vary with level of potassium and/or calcium)

Sulphur (S)

Atomic Number: 16   **Atomic Weight**: 32.06

Discoverer of Essentiality and Year: von Sachs, Knop—1865

Designated Element: Major element

Functions: Constituent of two amino acids, cystine and thiamine; component of compounds that give unique odor and taste to some types of plants

Mobility: Moderately mobile

Form Utilized: Sulfate (SO_4^{2-}) anion

Common Reagent Sources:

	Formula	% S
Ammonium sulfate	$(NH_4)_2SO_4$	24 (21% N)
Calcium sulfate	$CaSO_4 \cdot 2H_2O$	23 (26% Ca)
Magnesium sulfate	$MgSO_4 \cdot 7H_2O$	23 (10% Mg)
Potassium sulfate	K_2SO_4	17 (42% K)

Concentration in Nutrient Solutions: 70–150 mg/L (ppm)

Typical Deficiency Symptoms: General loss of green color of the entire plant; slowed growth

Symptoms of Excess: Not well defined

Critical Level: 0.30%

Sufficiency Ranges: 0.4–1.0%; cucumber, fully developed leaf: 0.4–0.7%

Excessive Level: Not known

Boron (B)

Atomic Number: 5   **Atomic Weight**: 10.81

Discoverer of Essentiality and Year: Sommer and Lipman—1926

Designated Element: Micronutrient

Functions: Associated with carbohydrate chemistry, pollen germination, and cellular activities (division, differentiation, maturation, respiration, and growth); important in the synthesis of one of the bases for RNA formation

Mobility: Immobile

Forms Utilized: Borate (BO_3^{3-}) anion as well as the molecule H_3BO_3

Common Reagent Sources:

	Formula	% B
Boric acid	H_3BO_3	16

Concentration in Nutrient Solutions: 0.3 mg/L (ppm)

Typical Deficiency Symptoms: Slowed and stunted new growth, with possible death of the growing point and root tips; lack of fruit set and development

Symptoms of Excess: Accumulates in the leaf margins, resulting in death of the margins

Critical Level: 25 mg/kg (ppm)

Sufficiency Ranges:

Plant	Plant Part	mg/kg (ppm) B
Head lettuce	Whole head	22–65
Lettuce	Wrapper leaf	25–60
Bell pepper	Fully developed leaf	25–75
Tomato	Opposite below flower cluster	
	First cluster	25–75
	Second cluster	25–75
	Third cluster	25–75
	Fourth cluster	25–75
	Fifth cluster	25–75
	Sixth cluster	25–75
Cucumber	Fully developed leaf	30–100

Toxic Level: >100 mg/kg (ppm)

Chlorine (Cl)

Atomic Number: 17   **Atomic Weight**: 35.45

Discoverer of Essentiality and Year: Stout—1954

Designated Element: Micronutrient

Functions: Involved in the evolution of oxygen (O_2) in photosystem II; raises cell osmotic pressure and affects stomatal regulation; increases hydration of plant tissue

Mobility: Mobile

Form Utilized: Chloride (Cl^-) anion

Common Reagent Sources:

	Formula	% Cl
Potassium chloride	KCl	47 (50% K)

Concentration in Nutrient Solutions: 50–1000 mg/L (ppm) (depends on reagents used)

Typical Deficiency Symptoms: Chlorosis of the younger leaves; wilting

Symptoms of Excess: Premature yellowing of leaves; burning of leaf tips and margins; bronzing and abscission of leaves

Critical Level: 20 mg/kg (ppm)

Sufficiency Range: 20–1500 mg/kg (ppm)

Excess Level: >0.50%

Copper (Cu)

Atomic Number: 29   **Atomic Weight**: 64.54

Discoverer of Essentiality and Year: Sommer—1931

Designated Element: Micronutrient

Functions: Constituent of the chloroplast protein plastocyanin; participates in electron transport system linking photosystem I and II; participates in carbohydrate metabolism and nitrogen (N_2) fixation

Mobility: Immobile

Form Utilized: Cupric (Cu^{2+}) cation

Common Reagent Sources:

	Formula	% Cu
Copper sulfate	$CuSO_4 \cdot 5H_2O$	25 (13% S)

Concentration in Nutrient Solutions: 0.01–0.1 mg/L (ppm); highly toxic to roots when in excess of 1.0 mg/L (ppm) in solution

Typical Deficiency Symptoms: Reduced or stunted growth, with a distortion of the young leaves; necrosis of the apical meristem

Symptoms of Excess: Induced iron deficiency and chlorosis; root growth will cease and root tips will die and turn black

Critical Level: 5 mg/kg (ppm)

Sufficiency Ranges:

Plant	Plant Part	mg/kg (ppm) Cu
Head lettuce	Whole head	5–17
Lettuce	Wrapper leaf	8–25
Bell pepper	Fully developed leaf	6–25
Tomato	Opposite below flower cluster	
	First cluster	5–50
	Second cluster	5–50
	Third cluster	5–50
	Fourth cluster	5–50
	Fifth cluster	5–50
	Sixth cluster	5–50
Cucumber	Fully developed leaf	8–10

Toxic Level: >50 mg/kg (ppm)

Iron (Fe)

Atomic Number: 26   **Atomic Weight**: 55.85

Discoverer of Essentiality and Year: von Sachs, Knop—1860

Designated Element: Micronutrient

Functions: Component of many enzyme and electron transport systems; component of protein ferredoxin; required for nitrate (NO_3) and sulfate (SO_4) reduction, nitrogen (N_2) assimilation, and energy (NADP) production; associated with chlorophyll formation

Mobility: Immobile

Forms Utilized: Ferrous (Fe^{2+}) and ferric (Fe^{3+}) cations

Common Reagent Sources:

	Formula	% Fe
Iron chelate	FeEDTA	6–12
Iron citrate	—	
Iron tartrate	—	
Iron lignin sulfonate	—	6
Ferrous sulfate	$FeSO_4 \cdot 7H_2O$	20 (11% S)

Concentration in Nutrient Solutions: 2–12 mg/L (ppm)

Typical Deficiency Symptoms: Interveinal chlorosis of younger leaves; as deficiency intensifies, older leaves are affected and younger leaves turn yellow

Symptoms of Excess: Not known for crops commonly grown hydro-ponically

Critical Level: 50 mg/kg (ppm)

Sufficiency Ranges:

Plant	Plant Part	mg/kg (ppm) Fe
Head lettuce	Whole head	56–560
Lettuce	Wrapper leaf	50–100
Bell pepper	Fully developed leaf	60–300
Tomato	Opposite below flower cluster	
	First cluster	60–300
	Second cluster	60–300
	Third cluster	60–300
	Fourth cluster	60–300
	Fifth cluster	60–300
	Sixth cluster	60–300
Cucumber	Fully developed leaf	50–300

Excess Level: Not known

Manganese (Mn)

Atomic Number: 25   **Atomic Weight**: 54.94

Discoverer of Essentiality and Year: McHargue—1922

Designated Element: Micronutrient

Functions: Involved in oxidation-reduction processes in the photosynthetic electron transport system; photosystem II for photolysis; activates IAA oxidases

Mobility: Immobile

Form Utilized: Manganous (Mn^{2+}) cation

Common Reagent Sources:

	Formula	% Mn
Manganese sulfate	$MnSO_4 \cdot 4H_2O$	24 (14% S)

Concentration in Nutrient Solutions: 0.5–2.0 mg/L (ppm)

Typical Deficiency Symptoms: Reduced and stunted growth, with interveinal chlorosis on younger leaves

Symptoms of Excess: Older leaves show brown spots surrounded by chlorotic zone or circle; black spots will appear on stems and petioles

Critical Level: 25 mg/kg (ppm)

Sufficiency Ranges:

Plant	Plant Part	mg/kg (ppm) Mn
Head lettuce	Whole head	30–200
Lettuce	Wrapper leaf	15–250
Bell pepper	Fully developed leaf	50–250
Tomato	Opposite below flower cluster	
	First cluster	50–250
	Second cluster	50–250
	Third cluster	50–250
	Fourth cluster	50–250
	Fifth cluster	50–250
	Sixth cluster	50–250
Cucumber	Fully developed leaf	30–300

Toxic Level: >400 mg/kg (ppm)

Molybdenum (Mo)

Atomic Number: 42   **Atomic Weight**: 95.94

Discoverer of Essentiality and Year: Sommer and Lipman—1926

Designated Element: Micronutrient

Functions: Component of two enzyme systems, nitrogenase and nitrate reductase, for the conversion of nitrate (NO_3) to ammonium (NH_4)

Mobility in Plant: Immobile

Form Utilized: Molybdate (MoO_4^{2-}) anion

Common Reagent Sources:

	Formula	% Mo
Ammonium molybdate	$(NH_4)_6Mo_7O_{24} \cdot 4H_2O$	8 (1% N)

Concentration in Nutrient Solutions: 0.05 mg/L (ppm)

Typical Deficiency Symptoms: Resemble nitrogen deficiency symptoms, with older and middle leaves becoming chlorotic; leaf margins will roll; growth and flower formation restricted

Symptoms of Excess: Not known

Critical Level: Not exactly known, probably 0.10 mg/kg (ppm)

Sufficiency Range: 0.2–1.0 mg/kg (ppm)

Excess Level: Not known

Zinc

Atomic Number: 30   **Atomic Weight**: 65.39

Discoverer of Essentiality and Year: Lipman and MacKinnon—1931

Designated Element: Micronutrient

Functions: Involved in same enzymatic functions as manganese and magnesium; specific to the enzyme carbonic anhydrase

Mobility: Immobile

Form Utilized: Zinc (Zn^{2+}) cation

Common Reagent Sources:

	Formula	% Zn
Zinc sulfate	$ZnSO_4 \cdot 7H_2O$	22 (11% S)

Concentration in Nutrient Solutions: 0.05 mg/L (ppm); can be highly toxic to roots when in excess of 0.5 mg/L (ppm)

Typical Deficiency Symptoms: Upper new leaves will curl with rosette appearance; chlorosis in the interveinal areas of new leaves produces a banding effect; leaves will die and fall off; flowers will abscise

Symptoms of Excess: Plants may develop typical iron deficiency symptoms: chlorosis of young leaves

Critical Level in Plants: 15 mg/kg (ppm)

Sufficiency Ranges

Plant	Plant Part	mg/kg (ppm) Zn
Head lettuce	Whole head	33–196
Lettuce	Wrapper leaf	25–250
Bell pepper	Fully developed leaf	20–200
Tomato	Opposite below flower cluster	
	First cluster	20–200
	Second cluster	20–200
	Third cluster	20–200
	Fourth cluster	20–200
	Fifth cluster	20–200
	Sixth cluster	20–200
Cucumber	Fully developed leaf	25–200

Excess Level: >300 mg/kg (ppm)

Appendix C: Reference Books, Bulletins, and Videos

This appendix, while not an exhaustive listing, includes reference materials that are readily available and reflect the current state of the art for hydroponic/soilless growing.

Reference Books

Ball, V. (Ed.). 1985. *Ball Red Book: Greenhouse Growing,* 14th edition. Reston Publishing, Reston, VA.

Barber, S.A. 1995. *Soil Nutrient Bioavailability: A Mechanistic Approach,* 2nd edition. John Wiley & Sons, New York.

Barber, S.A. and D.R. Bouldin (Eds.). 1984. *Roots, Nutrient and Water Influx, and Plant Growth.* ASA Special Publication 136. American Society of Agronomy, Madison, WI.

Bennet, W.F. (Ed.). 1993. *Nutrient Deficiencies and Toxicities in Crop Plants.* APS Press, The American Phytopathological Society, St. Paul, MN.

Bentley, M. 1974. *Hydroponics Plus.* O'Conner Printers, Sioux Falls, SD.

Blanc, D. 1985. *Growing Without Soil.* Institut National de la Recherche Agronomique, Paris, France.

Bridwell, R. 1990. *Hydroponic Gardening,* revised edition. Woodbridge Press, Santa Barbara, CA.

Bunt, A.C. 1988. *Media and Mixes for Container-Grown Plants,* 2nd edition. Unwin Hyman, London, England.

Carson, E.W. (Ed.). 1974. *The Plant Root and Its Environment.* University Press of Virginia, Charlottesville, VA.

Cooper, A. 1976. *Nutrient Film Technique for Growing Crops.* Grower Books, London, England.

Cooper, A. 1979. *Commercial Applications of NFT.* Grower Books, London, England.

Cooper, A. 1988. *The ABC of NFT.* Grower Books, London, England.

Cooper, A. 1996. *The ABC of NFT, Nutrient Film Technique.* Casper Publications, Narrabeen, Australia.

Dalton, L. and R. Smith. 1991. *Hydroponic Gardening.* Cobb/Horwood Publications, Auckland, New Zealand.

Day, D. 1991. *Growing in Perlite.* Grower Digest 12. Grower Books, London, England.

Douglas, J.S. 1975. *Hydroponics: The Bengal System with Notes on Other Methods of Soilless Cultivation,* 5th edition. Oxford University Press, London, England.

Douglas, J.S. 1984. *Beginner's Guide to Hydroponics,* new edition. Pelham Books, London, England.

Douglas, J.S. 1985. *Advanced Guide to Hydroponics,* new edition. Pelham Books, London, England.

Ellis, C. and M.W. Swaney. 1974. *Soilless Growth of Plants,* 2nd edition. Nostrand Reinhold Company, New York.

Epstein, E. 1972. *Mineral Nutrition of Plants: Principles and Perspectives.* John Wiley & Sons, New York.

Gericke, W.F. 1940. *The Complete Guide to Soilless Gardening.* Prentice Hall, New York.

Gooze, J. 1986. *The Hydroponic Workbook: A Guide to Soilless Gardening.* Rocky Top Publishers, Stamford, NY.

Handreck, H.A. and N.D. Black. 1993. *Growing Media for Ornamental Plants and Turf,* 4th edition. University of New South Wales Press, Sydney, Australia.

Harris, D. 1988. *Hydroponics: The Complete Guide to Gardening Without Soil. A Practical Handbook for Beginners, Hobbyists and Commercial Growers.* New Holland Publishers, London, England.

Hewitt, E.J. 1966. *Sand and Water Culture Methods in Plant Nutrition.* Commonwealth Agricultural Bureaux, Bucks, England.

Hollis, H.F. 1964. *Profitable Growing Without Soil.* The English University Press, London, England.

Hudson, J. 1975. *Hydroponic Greenhouse Gardening.* National Graphics, Garden Grove, CA.

Hurd, R.G., P. Adams, D.M. Massey, and D. Price (Eds.). 1980. *Symposium on Research on Recirculating Water Culture.* Acta Horticulture No. 98. The Hague, The Netherlands.

Jones, J.B., Jr. 1983. *A Guide for the Hydroponic and Soilless Culture Grower.* Timber Press, Portland, OR

Jones, L., P. Beardsley, and C. Beardsley. 1990. *Home Hydroponics...and How to Do It!* revised edition. Crown Publishers, New York.

Kenyon, S. 1993. *Hydroponics for the Home Gardener,* revised edition. Key Porter Books, London, England.

Kramer, J. 1976. *Gardens Without Soil. House Plants, Vegetables, and Flowers.* Scribner, New York.

Lindsay, W.L. 1979. *Chemical Equilibria in Soils.* John Wiley & Sons, New York.

Lorenz, O.A. and D.N. Maynard. 1988. *Knott's Handbook for Vegetable Growers,* 3rd edition. John Wiley & Sons, New York.

Ludwick, A.E. (Ed.). 1990. *Western Fertilizer Handbook—Horticultural Edition.* Interstate Publishers, Danville, IL.

Maas, E.F. and R.M. Adamson. 1971. *Soilless Culture of Commercial Greenhouse Tomatoes.* Publication 1460. Information Director, Canada Department of Agriculture, Ottawa, Ontario, Canada.

Marshner, H. 1986. *Mineral Nutrition of Higher Plants.* Academic Press, New York.

Mason, J. 1990. *Commercial Hydroponics.* Kangaroo Press, Kenthurst, NSW, Australia.

Mengel, K. and E.A. Kirkby. 1987. *Principles of Plant Nutrition,* 4th edition. International Potash Institute, Worblaufen-Bern, Switzerland.

Mills, H.A. and Jones, J.B., Jr. 1996. *Plant Analysis Handbook II.* Micro-Macro Publishing, Athens, GA.

Mittleider, J.R. 1982. *More Food from Your Garden.* Woodbridge Press, Santa Barbara, CA.

Molyneux, C.J. 1988. *A Practical Guide to NFT.* T. Snap & Co. Ltd., Preston, Lancashire, England.

Muckle, M.E. 1982. *Basic Hydroponics.* Growers Press, Princeton, B.C., Canada.

Muckle, M.E. 1990. *Hydroponic Nutrients—Easy Ways to Make Your Own,* revised edition. Growers Press, Princeton, B.C., Canada.

Nicholls, R.E. 1990. *Beginning Hydroponics: Soilless Gardening: A Beginner's Guide to Growing Vegetables, House Plants, Flowers, and Herbs Without Soil.* Running Press, Philadelphia, PA.

Pais, I. and J.B. Jones, Jr. 1997. *The Handbook of Trace Elements.* St. Lucie Press, Boca Raton, FL.

Parker, D. (Ed.). 1984. *The Best of Growing Edge.* New Moon Publishing, Corvallis, OR.

Resh, H.M. 1990. *Hydroponic Home Food Gardens.* Woodbridge Press, Santa Barbara, CA.

Resh, H.M. 1993. *Hydroponic Tomatoes for the Home Gardener.* Woodbridge Press, Santa Barbara, CA.

Resh, H.M. 1995. *Hydroponic Food Production.* Woodbridge Press, Santa Barbara, CA.

Roorda van Eysinga, J.P.N.L. and K.W. Smith. 1981. *Nutritional Disorders in Glasshouse Tomatoes, Cucumbers, and Lettuce.* Centre for Agricultural Publishing and Documentation, Wageningen, The Netherlands.

Saunby, T. 1974. *Soilless Culture,* 3rd printing. Transatlantic Arts, Levittown, NY.

Savage, A.J. (Ed.). 1985. *Hydroponics Worldwide: State of the Art in Soilless Crop Production.* International Center for Special Studies, Honolulu, HI.

Savage, A.J. 1985. *Master Guide to Planning Profitable Hydroponic Greenhouse Operations,* revised edition. International Center for Special Studies, Honolulu, HI.

Savage, A.J. 1989. *Master Guide to Planning Profitable Hydroponic Greenhouse Operations,* revised edition. International Center for Special Studies, Honolulu, HI.

Scaife, A. and M. Turner. 1984. *Diagnosis of Mineral Disorders in Plants,* Volume 2: Vegetables. Chemical Publishing Company, New York.

Schwarz, M. 1968. *Guide to Commercial Hydroponics.* Israel University Press, Jerusalem, Israel.

Smith, D.L. 1987. *Rock Wool in Horticulture.* Grower Books, London, England.

Sundstrom, A.C. 1989. *Simple Hydroponics—for Australian and New Zealand Gardeners,* 3rd edition. South Yarra, Victoria, Australia.

Sutherland, S.K. 1986. *Hydroponics for Everyone.* Hyland House, South Yarra, Victoria, Australia.

Taylor, J.D. 1983. *Grow More Nutritious Vegetables Without Soil: New Organic Method of Hydroponics.* Parkside Press, Santa Ana, CA.

Van Patten, G.F. 1990. *Gardening: The Rockwool Book.* Van Patten Publishers, Portland, OR.

Wallace, A. 1971. *Regulation of the Micronutrient Status by Chelating Agents and Other Factors.* UCLA 34P51-33. Arthur Wallace, Los Angeles, CA.

Wallace, T. 1961. *The Diagnosis of Mineral Deficiencies in Plants,* 3rd edition. Chemical Publishers, New York.

Weir, R.G. and G.C. Cresswell. 1993. *Plant Nutritional Disorders. Vegetable Crops,* Volume 3. Florida Science Source, Lake Alfred, FL.

Whiting, A. 1985. *Lettuce from Eden: Hydroponic Growing Systems for Small or Large Greenhouses.* A Whiting, Cargill, Ontario, Canada.

Whitter, S.H. and S. Honma. 1979. *Greenhouse Tomatoes, Lettuce, and Cucumbers.* Michigan State University Press, East Lansing, MI.

Bulletins and Miscellaneous Publications

Baker, K.F. (Ed.). 1957. *The U.C. System for Producing Healthy Container-Grown Plants.* California Agricultural Experiment Station Manual 23. Berkeley, CA.

Broodley, J.W. and R. Sheldrake, Jr. 1972. *Cornell Peat-Lite Mixes for Commercial Plant Growing.* Information Bulletin 43. Cornell University, Ithaca, NY.

Butler, J.D. and N.F. Oebker. 1962. *Hydroponics as a Hobby Growing Plants Without Soil.* Extension Circular #844. University of Illinois, Extension Service in Agriculture and Home Economics, Urbana, IL.

Collins, W.L. and M.H. Jensen. 1983. *Hydroponic, A 1983 Technology Overview.* Environmental Laboratory, University of Arizona, Tucson, AZ.

Ellis, N.H., M. Jensen, J. Larsen, and N. Oebker. 1974. *Nutriculture Systems—Growing Plants Without Soil.* Bulletin No. 44. Purdue University, West Lafayette, IN.

Epstein, E. and B.A. Krantz. 1972. *Growing Plants in Solution Culture.* Publication AXT-196. University of California, Davis, CA.

Farnhand, D.S., R.F. Hasek, and J.L. Paul. 1985. *Water Quality.* Division of Agricultural Science, Leaflet 2995. University of California, Davis, CA.

Gerber, J.M. 1985. *Hydroponics. Horticulture Facts.* VC-19-8 (revision). University of Illinois at Urbana-Champaign, Urbana, IL.

Gilbert, H. 1979. *Hydroponic and Soilless Cultures, 1969–May 1979.* Quick Bibliography Series—National Agricultural Library (79-02). USDA, Beltsville, MD.

Gilbert, H. 1983. *Hydroponics/Nutrient Film Technique, 1979–1983* (Bibliography). Quick Bibliography Series—National Agricultural Library (83-31). USDA, Beltsville, MD.

Gilbert, H. 1984. *Hydroponics/Nutrient Film Technique.* Quick Bibliography Series—National Agricultural Library (84-56). USDA, Beltsville, MD.

Gilbert, H. 1985. *Hydroponics/Nutrient Film Technique: 1979–85.* Quick Bibliography Series—National Agricultural Library (86-22). USDA, Beltsville, MD.

Gilbert, H. 1987. *Hydroponics/Nutrient Film Technique: 1981–1986.* Quick Bibliography Series—National Agricultural Library (87-36). USDA, Beltsville, MD.

Gilbert, H. 1992. *Hydroponics/Nutrient Film Technique: 1983–1991.* Quick Bibliography Series—National Agricultural Library (92-43). USDA, Beltsville, MD.

Hoagland, D.R. and D.I. Arnon. 1950. *The Water-Culture Method for Growing Plants Without Soil.* Circular 347. Agricultural Experiment Station, University of California, Berkeley, CA.

Jensen, M.H. 1971. *The Use of Polyethylene Barriers Between Soil and Growing Medium in Greenhouse Vegetable Production.* Environmental Research Laboratory, University of Arizona, Tucson, AZ.

Johnson, H., Jr. 1977. *Hydroponics: A Guide to Soilless Culture Systems.* Leaflet 2947. University of California, Davis, CA.

Johnson, H., Jr. 1980. *Hydroponics: A Guide to Soilless Culture Systems.* Leaflet 2947. Division of Agricultural Sciences, University of California, Davis, CA.

Johnson, H., Jr. G.J. Hochmuth, and D.N. Maynard. 1985. *Soilless Culture of Greenhouse Vegetables.* IFAS Bulletin 218. C.M. Hinton Publications Distribution Center, Cooperative Extension Service, University of Florida, Gainesville, FL.

Jutras, M.N. 1979. *Nutrient Solutions for Plants (Hydroponic Solutions). Their Preparation and Use.* Circular 182. South Carolina Agricultural Experiment Station, Clemson, SC.

Kratky, B.A. 1996. *Non-Circulating Hydroponic Methods.* DPL Hawaii, Hilo, HI.

Larsen, J.E. 1971. *A Peat-Vermiculite Mix for Growing Transplants and Vegetables in Trough Culture.* Texas A&M University, College Station, TX.

Larsen, J.E. 1971. *Formulas for Growing Tomatoes by Nutriculture Methods (Hydroponics).* Mimeo. Texas A&M University, College Station, TX.

Maynard, D.N. and A.V. Baker. 1970. *Nutriculture—A Guide to the Soilless Culture of Plants.* Publication No. 41. University of Massachusetts, Amherst, MA.

Philipsen, D.J., J.L. Taylor, and I.E. Widders. 1985. *Hydroponics at Home.* Extension Bulletin E-1853. Cooperative Extension Service, Michigan State University, East Lansing, MI.

Schippers, P.A. 1977. *Annotated Bibliography on Nutrient Film Technique.* Vegetable Crops Mimeo 186. Cornell University, Ithaca, NY.

Schippers, P.A. 1977. *Construction and Operation of the Nutrient Flow Technique.* Vegetable Crops Mimeo 187. Cornell University, Ithaca, NY.

Schippers, P.A. 1979. *The Nutrient Flow Technique.* Vegetable Crops Mimeo 212. Department of Vegetable Crops, Cornell University, Ithaca, NY.

Sheldrake, R., Jr. and J.W. Boodley. 1965. *Commercial Production of Vegetable and Flower Plants.* Cornell Extension Bulletin 1065. Cornell University, Ithaca, NY.

Sonneveld, C. 1985. *A Method for Calculating the Composition of Nutrient Solutions for Soilless Cultures.* No. 10. Second translated edition. Glasshouse Crops Research and Experiment Station, Naaldwijk, The Netherlands.

Stoughton, R.H. 1969. *Soilless Cultivation and Its Application to Commercial Horticultural Crop Production.* Document No. MI/95768. Administration Unit, Distribution and Sales Section, FAO of the United Nations, via delle Terme de Caracalla, Rome, Italy.

Stout, J.G. and M.E. Marvel. 1966. *Hydroponic Culture of Vegetable Crops.* Circular 192-A. Florida Agricultural Extension Service, Gainesville, FL.

Taylor, J.D. and R.L. Flannery. 1970. *Growing Greenhouse Tomatoes in a Peat-Vermiculite Media.* Vegetable Crops Offsets Series #33. College of Agriculture and Environmental Science, Rutgers University, New Brunswick, NJ.

University of Arizona. 1973. *Annual Report, Environmental Research Laboratory, University of Arizona and Arid Lands Research Center, Abu Dhabi.* Tucson, AZ.

White, J.W. 1974. *Dillon Research Fund, Progress Report on Research at Penn State. Pennsylvania Flower Growers Bull.* 89:3-4. Pennsylvania State University, University Park, PA.

Wilcox, G.E. 1981. *Growing Greenhouse Tomatoes in the Nutrient Film Hydroponic System.* Bulletin 167. Purdue University Agricultural Experiment Station, West Lafayette, IN.

Wilcox, G.E. 1981. *The Nutrient Film Hydroponic System.* Bulletin 166. Purdue University Agricultural Experiment Station, West Lafayette, IN.

Wilcox, G.E. 1987. *NFT and Principles of Hydroponics.* Bulletin 530. Purdue University Agricultural Experiment Station, West Lafayette, IN.

Videos

Grower Application of Media and Tissue Analysis. 1993. St. Lucie Press, 2000 Corporate Blvd. NW, Boca Raton, FL 33431.

Grower Training Workshop on Video. 1995. CropKing, 5050 Greenwich Road, Seville, OH 44273.

Hobby Hydroponics. 1996. Nelson/Pade Multimedia, Maripola, CA 95338.

Hobby Hydroponics—Introduction. 1994. CropKing, 5050 Greenwich Road, Seville, OH 44273.

Hydroponic: An Introduction to Soilless Agriculture. 1990. American Association for Vocational Instructional Materials, 220 Smithsonia Road, Winterville, GA 30683.

Hydroponic Explained. 1993. Ausponics, P.O. Box 572, Sutherland, NSW 2232, Australia.

Hydroponic Gardening: Step-by-Step. 1992. Practical Hydroponics Magazine, P.O. Box 879, Bondi Junction, NSW 2022, Australia.

Hydro Systems Aqua Nutrient Growing System. 1997. ON SAT ONE, 301 Concord Road, Anderson, SC 29621.

Inside Hydroponics: The Inside Scoop on Inside Gardening. Hygro Technologies, 1972 Edmonds Way, Suite 206, Edmonds, WA 98020.

Introduction to Hydroponics. 1996. Superior Growers, 2819 Crow Canyon Road, Suite 218, San Ramon, CA 94583.

Non-Circulating Hydroponic Methods. 1995. University of Hawaii at Hilo, College of Agriculture, Hilo, HI 96720.

Nutrient Element Deficiencies in Tomato. 1993. St. Lucie Press, 2000 Corporate Blvd. NW, Boca Raton, FL 33431.

Nutrient Film Technique: The Easy Gardening Series, Volume II. 1993. Virginia Hydroponics, P.O. Box 7327, Hampton, VA 23666.

Your Future in Hydroponics. 1995. CropKing, 5050 Greenwich Road, Seville, OH 44273.

CD-ROM

The Encyclopedia of Hydroponic Gardening CD-ROM. 1996. Nelson/Pade Multimedia, Mariposa, CA 95338.

Appendix D:
How to Create
Nutrient Element
Deficiency Symptoms
Hydroponically:
A Student's Guide

Introduction

A common school science-fair project demonstrates some form of hydro-ponic/soilless plant growing. The information in this appendix will be help-ful to a student contemplating such a project. The hydroponic procedures that a student can follow in order to generate nutrient element deficiency symptoms and monitor their effects on plant growth and development are described. With easily obtainable items and properly prepared nutrient so-lutions, the student should be able to undertake such a science project and obtain good results in about 6 to 8 weeks. The references in this section provide specific information that will be helpful to a student with this project.

Required Items

- One-liter plastic Coke bottles
- Horticultural perlite (available at most garden centers)
- 6″ × 6″ plastic refrigerator storage boxes with lids
- 50-mL graduated cylinder (available from most chemical apparatus supply houses)
- Green bean seeds (bush beans recommended)
- Pure water
- Nutrient solution reagents (obtainable from most chemical supply houses or hydroponic suppliers)

Reagents	Formula
Major Element Sources	
Calcium nitrate	$Ca(NO_3)_2 \cdot 4H_2O$
Calcium sulfate	$CaSO_4 \cdot 2H_2O$
Monocalcium phosphate	$Ca(H_2PO_4)_2 \cdot H_2O$
Potassium nitrate	KNO_3
Potassium sulfate	K_2SO_4
Monopotassium phosphate	KH_2PO_4
Magnesium sulfate	$MgSO_4 \cdot 7H_2O$
Magnesium nitrate	$Mg(NO_3)_2 \cdot 6H_2O$
Micronutrient Sources	
Boric acid	H_3BO_3
Copper sulfate	$CuSO_4 \cdot 5H_2O$
Manganese chloride	$MnCl_2 \cdot 4H_2O$
Manganese sulfate	$MnSO_4 \cdot H_2O$
Molybdic acid	$H_2MoO_4 \cdot H_2O$
Zinc sulfate	$ZnSO_4 \cdot H_2O$
Iron chelate	FeNaEDTA

Growing Requirements

Light—Plant growth is best when plants are exposed to full sunlight for at least 8 hours each day. Placing plants by a window, even one that is well lit, is not sufficient for best growth. Use of lights to extend the exposure time is not an adequate substitute for natural sunlight. Slow growth and development, usually seen as spindly-looking plants, are signs of inadequate light. If a nutrient element deficiency symptom is to be developed, plant exposure to full sunlight is required for success.

Plant Species Selection—For best results in a reasonable length of time, a plant species that is rapid growing and responsive to its environment should be selected. Experience has shown that green bean is probably the best plant species, and corn is second best. Although other plant species are faster growing, such as radish and lettuce, the larger plant size of the green bean and corn plants makes them the best choices.

Temperature—Plants grow best when the air temperature is maintained between 75° to 85°F (24 to 30°C). Air temperatures above or below these limits are not conducive to normal plant growth and development.

Moisture—Plants that are cycled through periods of adequate and then inadequate water supply will develop abnormal growth appearances due to that stress. Therefore, plants must have access to an adequate supply of water at all times. However, overwatering is as detrimental to plant growth and development as inadequate watering. Frequent small doses of water added to the rooting medium are better than infrequent heavy doses. The growing technique described in this section will maintain an adequate water supply for the plants at all times.

Pest Control—Insects and disease problems can be avoided by keeping the growing area clean at all times and free from potential sources of infestation. Although neighboring plants may be free from visual pests, it is wise to conduct the experiments given in this section free from the presence of other plants that are not part of the study.

Procedure

1. Remove the top of a one-liter plastic Coke bottle by cutting around the bottle at the upper label level (see Figure 26). The number of bottles needed will depend on which experiments will be conducted. Only one bottle per treatment is needed, although duplicates will ensure that there will be a backup treatment bottle(s) if one bottle is lost. Also, one bottle plus its backup will be needed as the *check* (that without a treatment change).
2. Drill a ¹/₂-inch-diameter hole in the center of the bottom of the bottle.

Figure 26   A one-liter Coke bottle with its top removed.

3. On the inside of the bottle, cover the hole in the bottom of the bottle with plastic mesh. The plastic mesh will prevent the lost of perlite from the opening in the bottle.
4. Fill the bottle with horticultural-grade perlite all the way to the top.
5. Using the prepared nutrient solution, leach the perlite until the solution freely flows from the hole in the bottom of the bottle. The plastic mesh, if properly in place, will keep the perlite from being lost from the bottom of the bottle.
6. Place two green bean seeds about an inch deep into the moist perlite. It may be necessary to add a small amount of water (half a cup) daily to the top of the bottle to keep the perlite moist until the seeds germinate and the cotyledons appear.
7. Place the bottle into the small plastic refrigerator box and fill the box with 2 inches of nutrient solution. With a black marker pen, put a scribe mark at the nutrient solution level on the side of the refrigerator box. When adding nutrient solution, always fill to that mark. Cut an opening in the box lid large enough to just accommodate the bottle. Place the refrigerator lid on the box and snap down tight (see Figure 27). Keeping the lid in place will prevent evaporation of the

Figure 27   The one-liter Coke bottle set in the refrigerator box.

nutrient solution. The nutrient solution in the box will also fill the bottle with nutrient solution at that same level.

8.   Place the bottle in its box in full sunlight. Add nutrient solution when needed (usually every day) to maintain the level in the box (at the scribe mark) using a 50-mL graduated cylinder so that water use can be monitored.

9.   When the seeds germinate, remove one of the seedlings to leave just one plant per bottle.

10.   When the plants reach the two-leaf stage, begin the treatments.

Nutrient Element Deficiency Experiments

Introduction

Visual nutrient deficiency symptoms for the major elements (calcium, magnesium, nitrogen, phosphorus, and potassium) are fairly easy to develop using the technique that is to be described. Deficiencies of the micronutrients (boron, chlorine, copper, iron, manganese, molybdenum, and zinc) are more challenging and difficult to achieve. The reason is that the major elements are required in substantial quantities by plants, whereas the micronutrients are not. It is quite difficult to deplete the growing medium and the nutrient solution of trace quantities of the micronutrient elements necessary to create a deficient condition. In addition, there may be a sufficient quantity of a micronutrient in the plant itself to satisfy the requirement until the plant reaches full maturity. However, it may be worth a try if you like a challenge.

Upon reaching Step 10 in the procedure list, the composition of the nutrient solution is altered to free it of one of the essential elements, as shown in Table 68.

Table 68   Preparation of Hoagland nutrient solutions for nutrient deficiency symptom development

Stock Solution (g/L)	mL Stock Solution per Liter Nutrient Solution							
	Complete	-N	-P	-K	-Ca	-Mg	-S	-Fe
1 M Ca(NO$_3$)$_2$ · 4H$_2$O (236)	5	—	4	5	—	4	4	5
1 M KNO$_3$ (101)	5	—	6	—	5	6	6	5
1 M KH$_2$PO$_4$ (136)	1	—	—	—	1	1	1	1
1 M MgSO$_4$ · 7H$_2$O (246)	2	2	2	2	2	—	—	2
50 mM FeNaEDTA (18.4)*	1	1	1	1	1	1	1	—
Micronutrients**	1	1	1	1	1	1	1	1
0.05 M K$_2$SO$_4$ (87)	—	5	—	—	—	3	—	—
0.01 M CaSO$_4$ · 2H$_2$O (1.72)	—	200	—	—	—	—	—	—
0.05 M Ca(H$_2$PO$_4$)$_2$ · H$_2$O (12.6)	10	—	10	—	—	—	—	—
1 M Mg(NO$_3$)$_2$ · 6H$_2$O (256)	—	—	—	—	—	—	2	—

 * Ferric-sodium salt of ethylenediaminetetraacetic acid (EDTA). Differs from Hoagland recipe, which uses iron tartrate.
** Contains the following:

g/L	Reagent
2.86	H$_3$BO$_3$
1.18	MnCl$_2$ · 4H$_2$O
0.22	ZnSO$_4$ · 7H$_2$O
0.08	CuSO$_4$ · 5H$_2$O
0.02	H$_2$MoO$_4$ · H$_2$O (85% molybdic acid)

Source: Hoagland and Arnon, 1950.

For boron-, copper-, manganese-, molybdenum-, and zinc-deficient solutions, substitute micronutrient stock solutions for one of the five salts in the regular micronutrient stock solution. For chlorine-deficient micronutrient solution, substitute 1.55 $MnSO_4 \cdot H_2O$ for 1.18 $MnCl_2 \cdot 2H_2O$.

Procedure

1. Remove the Coke bottle from the plastic box and leach the bottle with pure water until there is a free flow of water from the hole in the bottom of the bottle. This leaching procedure will free the perlite from any accumulated nutrient solution in the bottle.
2. When the flow of water from the bottom of the bottle ceases, place the bottle into the refrigerator box containing one of the treatment nutrient solutions derived from Table 68 (a nutrient solution free from one of the major elements). Be sure to keep at least one bottle on the "full" treatment so that a visual comparison can be made between the plant receiving all of the essential elements versus those missing one of the essential elements.
3. Place the bottle back into its refrigerator box and then back into full sunlight.
4. Maintain the nutrient solution level in the box by adding nutrient solution periodically (usually daily) as required, and record the milliliters of solution required to bring back to the scribe mark on the side of the box.
5. Depending on the light conditions and rate of growth, significant changes in plant appearance should become evident in about 10 days to 2 weeks.
6. The first evidence of deficiency will be slowed growth.

Photographic Record

It would be useful to have a photographic record of the plants at critical stages of development. A daily record can be expensive in terms of the film required but highly useful in observing the change in plant appearance with time. In order to obtain a meaningful visual record, plant and camera placement is critical. A simple backdrop, called a studio box, can be constructed from a large cardboard box and a piece of blue burlap cloth.

Cut the cardboard box on the diagonal, and line the inside of the box with blue burlap cloth on the bottom and up the inside of the box, cutting to just fit the inside box opening Take one of the box-bottle containers and place it in the center of the bottom of the cardboard box. Draw a square around the box-bottle container, which will designate where the box-bottle should be placed each time a photograph is to be taken. Be sure to also place a mark on the side of the box-bottle container so that it is always oriented in the same way when the photograph is taken.

On the back side of the box, cut small holes just on the inside edge of the back side of the box at 2-inch intervals. Using thick white string, pull a length of string through the holes. Placing white lines of string at 2-inch intervals up the inside back of the studio box will provide a 2-inch measuring backdrop. A picture of such a constructed box and a box-bottle container in place is shown in Figure 28.

Figure 28   Box-bottle set in the studio box for photographing.

For those who have access to a video camcorder, a similar photographic record can be made using daily short exposures of the plants placed in the studio box; be certain that each day's exposure is exactly positioned (video camera and plant). The short daily exposures can then be edited to give a time-lapse record of the plant as the deficiency symptoms develop.

Plant Growth Record

A daily record should be kept, observing water use and plant growth. Height measurements may be of little value, since, for example, the development of lateral branches is the primary indicator of plant growth for green bean, whereas plant height would be the proper measurement for corn. As the deficiency develops, changes in plant growth will also be influenced by the environmental conditions, such as light and temperature. The interaction between these environmental factors and the developing deficiency symptoms can make for an interesting study.

References

The student will find the following references useful in conducting hydroponic studies:

Brooke, L.L. and O. Silberstein. 1993. Hydroponic in schools: An educational tool. *The Growing Edge* 5:20–22, 66.

Hershey, D.R. 1992. Inexpensive hydroponic teaching methods, pp. 27–34. In: D. Schact (Ed.), *Proceedings of the 13th Annual Conference on Hydroponics*. Hydroponic Society of America, San Ramon, CA.

Hershey, D.R. 1995. *Plant Biology Science Projects*. John Wiley & Sons, New York.

Hershey, D.R. and G.W. Stutte, 1991. A laboratory exercise on semiquantitative analysis of ions in nutrient solution. *J. Agron. Educ.* 20:7–10.

Jones, J.B., Jr. 1985. Growing plants hydroponically. *Am. Biol. Teacher* 47:356–358.

Lopez, L.M. 1981. Hydroponics—Studies in plant culture with historical roots. *Sci. Teacher* 48(6):47–49.

Nicol, E. 1990. Hydroponics and aquaculture in the high school classroom. *Am. Biol. Teacher* 52:182–184.

Parker, D. 1993. Systems for beginners—Hydro 101. *The Growing Edge* 5:61–66.

Victorian Schools' Nursery. 1986. *Hydroponics for Schools and the Home Grower.* Victorian Schools' Nursery, Melbourne, Australia.

Warner, P.A., D.A. Rakow, and C. Mazza. 1993. *Grow with the Flow: A Hydroponic Gardening Project.* Leader's/Members' Guide 141M7. Cornell Cooperative Extension Service, Cornell University, Ithaca, NY.

Index